Modern Project Management Techniques

for the

Environmental Remediation Industry

Timothy J. Havranek, PMP

CRC Press
Taylor & Francis Group
Boca Raton   London   New York

CRC Press is an imprint of the
Taylor & Francis Group, an **informa** business

CRC Press
Taylor & Francis Group
6000 Broken Sound Parkway NW, Suite 300
Boca Raton, FL 33487-2742

First issued in paperback 2019

© 1999 by Taylor & Francis Group, LLC
CRC Press is an imprint of Taylor & Francis Group, an Informa business

No claim to original U.S. Government works

ISBN-13: 978-1-57444-218-2 (hbk)
ISBN-13: 978-0-367-40021-7 (pbk)

Library of Congress Cataloging-in-Publication Data

Catalog information may be obtained from the Library of Congress

**Visit the Taylor & Francis Web site at
http://www.taylorandfrancis.com**

**and the CRC Press Web site at
http://www.crcpress.com**

Table of Contents

To my parents,
Mary Jane Havranek and John Francis Havranek

Preface

Since the mid-1970s, industrial/commercial site owners and environmental service companies have been working to confront the technical challenges posed by environmental remediation projects. Although meeting these challenges has been difficult, the industry can boast much success in the development of new, more effective remediation technologies and sophisticated site investigation techniques. In addition, the various environmental regulations that have evolved over the years in many cases are becoming more realistic and pragmatic.

Despite technical successes and positive regulatory changes, the cost to achieve cleanup at any given site continues to remain high and the time required to achieve site closure, in most cases, remains quite long. Over the years, environmental remediation projects have become notorious for their high budget overruns and long schedule delays. The industry has matured and many are beginning to suspect (and rightly so) that the solution to cost-effective and timely project completion may be found more in the proper management of technology implementation than in the technology itself.

Environmental remediation is not the domain of any particular technical discipline or profession. A skilled, properly integrated, multidisciplined team, cognizant of the various technical, political, economic, and regulatory constraints, is required for cost-effective, timely, and high-quality project completion. It is the role of the project manager to see to it that the responding team is properly integrated and focused. This requires the utilization of a full gamut of planning, communication, and monitoring techniques as presented throughout this book.

Environmental service companies experienced a tremendous growth boom in the 1980s and early 1990s. During the mid-1990s, this growth boom was replaced by increased competition and reduced profits. This type of trend is common in all industries and it is at the point of increased competition, and reduced profits, that project management becomes of importance to an industry.

Most environmental service companies are in search of new ways to survive and prosper in the new market. These companies are in dire need of modern project management processes.

The overall objective of this book is to provide a means for readers to substantially improve remediation project performance through the implementation of project management techniques geared specifically to the environmental remediation industry. With this objective in mind, this book attempts to pragmatically apply the best of modern project management to the issues of environmental remediation. Its focus is on the tools and techniques for implementing project management processes. Its theme is one of optimization, in terms of the amount of money, time, energy, and other resources expended in completing these projects. As a management book, it does not address the detailed issues of environmental assessment, remediation system design, or regulatory interpretation. With little effort, the reader can find a number of books and articles pertaining to those issues.

The intended audience is anyone responsible for the management of environmental remediation projects or interested in improving the overall efficiency of such projects. Those involved at the project management level and above (mid-level and senior level managers) in environmental service companies may find it especially helpful as it deals with issues of customer relations, cost-effective project completion, and project profitability. Environmental managers and legal affairs groups within industrial corporations, faced not only with managing environmental remediation projects but with overseeing the efforts of environmental service companies, will find strategies for reducing project costs, preparing for uncertainties, and replacing reactive remediation planning with proactive, sound business management. Academic institutions, especially those that wish to teach project management from the viewpoint of application to a specific industry, may find the book useful in graduate-level and continuing education courses.

Finally, it is my belief that if the project management techniques presented in this book were to become part of the overall culture of managing environmental remediation projects, the benefits will accrue not only to those organizations making use of the techniques, but to the environment and the economy as well.

Acknowledgments

I would like to express special thanks to the following individuals who played significant roles in the development of this book: Jamie Z. Carlson, who provided graphics development and administrative assistance. Without Jamie's dedicated help and persistence, this project may never have been completed. Wendy Cross, for her assistance in editing original text and providing overall writing encouragement. Also thanks to William M. Lavin and Michael H. Sullivan, who provided the original opportunity to present these concepts in a training format.

PROJECT MANAGEMENT INSTITUTE

The author expresses appreciation to the Project Management Institute (PMI) for various materials used throughout this book. A nonprofit professional association, PMI represents members worldwide who are actively advancing the project management profession. This prestigious institute is recognized by the profession internationally as the official certifying organization for the Project Management Professional (PMP®).

PMI establishes project management standards and provides seminars, educational programs, and professional certification that organizations desire for their project leaders. Members are kept well informed with two highly respected, professional magazines, *Project Management Journal* and *PM Network.*

Since its founding in 1969, PMI has grown to be the organization of choice for top project professionals in a wide variety of disciplines. PMI's *A Guide to the Project Management Body of Knowledge,* Education Programs, and PMP® Certification aid members in developing their project management skills and careers. PMP® certification is increasingly viewed as an indication of valuable knowledge by many industries and is recognized worldwide by major corporations.

For more information, contact PMI at Four Campus Boulevard, Newtown Square, PA 19073-3299 USA; phone: (610) 356-4600; fax: (610) 356-4647; Website: www.pmi.org.

About the Author

Timothy J. Havranek, a certified Project Management Professional (PMP), is the founder and president of Environmental Project Management, Inc. (EPM), an environmental firm located near Pittsburgh, Pennsylvania. Mr. Havranek established EPM in February of 1994, to assist industry in reducing the cost and duration of environmental remediation projects through the practical application of modern project management theory. He holds a B.S. degree in Petroleum Engineering from Marietta College, Marietta, Ohio.

Mr. Havranek has managed more than 45 environmental projects, working with corporate site owners, regulatory agencies, and environmental remediation firms. His oversight and consulting services include advanced project planning using quantitative decision analysis and stochastic project modeling, identification of long-term/short-term goals and strategies, coordination of environmental planning groups, formal implementation of modern project management processes, and project management training. In the course of his career, Mr. Havranek has managed and coordinated the activities of geologists, engineers, toxicologists, environmental scientists, field technicians, and numerous subcontractors. His technical expertise includes contaminant delineation, fate and transport, groundwater and soil treatment (especially *in situ* and on-site), feasibility studies, regulatory negotiation, strategies for brownfields redevelopment, and environmental risk management.

Mr. Havranek has taught seminars such as "Cost Effective Environmental Project Management" and "Project Planning for Environmental Remediation." He has over 16 years of experience as an environmental project manager/engineer and offshore construction manager in the oil and gas exploration industry. An active member of the Project Management Institute (PMI), he has served as the Pittsburgh Chapter chairman for professional development (1993) and since 1992 has served as a trainer in risk management for the PMP exam review course offered by the Pittsburgh Chapter of the PMI. Mr. Havranek is also a member of the AACE International.

Mr. Havranek was greatly inspired by the work of R. Buckminster Fuller, in particular with regard to design science and strategic planning. His research interests include quantitative decision analysis, general systems modeling, applied mathematics, and synergetic geometry.

Introduction

The term environmental remediation refers to the performance of engineering/construction projects designed to remedy or restore environmental media, particularly soil and groundwater, degraded by chemical compounds (or elements) that may pose a threat to human health and the environment. Most companies that manufacture a product or provide a service handle chemicals or produce waste materials which, under certain conditions, may pose a threat to human health and the environment. The various ways by which hazardous chemical products or waste materials have entered the environment include:

- Accidental releases or spills during loading, transportation, and unloading operations
- Improper handling, storage, and disposal at times when the potential hazards associated with various materials were not well understood
- Leaks from aging or deteriorated equipment such as underground storage tanks and their associated piping
- Intentional release due to unsophisticated or unethical business practices

The discovery of sites such as New York's Love Canal and Missouri's Times Beach in the late 1970s and early 1980s raised public awareness and concern about hazardous waste sites to an all-time high. The federal government responded with the development of new statutes such as the Comprehensive Environmental Response, Compensation, and Liability Act (CERCLA) (also known as Superfund) and enhanced or increased enforcement of regulations already in place under the Resource Conservation and Recovery Act (RCRA), the Toxic Substances Control Act (TSCA), and the Clean Water Act. At the time that the CERCLA, RCRA, and TSCA regulations were developed, it is safe to say that few individuals (if any) truly understood:

- The number of sites that would be discovered
- The technical challenges to be faced
- The amount of money that would be spent
- The size of the industry that would be created

1.1 ECONOMIC IMPACT OF ENVIRONMENTAL REMEDIATION

As the 1980s wore on, public concern over the heath effects of hazardous waste sites was rivaled only by concern over the cost to clean them up. Billions of dollars were spent during the 1980s for remediation of hazardous waste sites, yet by the end of the decade only a small fraction of the known hazardous waste sites had been adequately addressed. Concern increased further as respected news media began to report in depth on the problem. For example, in May 1991, the *Wall Street Journal* reported that the future cost to clean up hazardous waste sites under the federal government's Superfund program could range between $600 billion and $1 trillion.[1] Superfund was initiated to provide government funding to clean up hazardous waste sites when the responsible parties could not be found or, if identified, could not pay for their portion of the cleanup.

Cost concerns were not isolated to the Superfund program. During the 1980s, privately owned and operated industrial facilities involved in the treatment, storage, and disposal of hazardous chemicals were commonly found to have confirmed releases of chemicals to the subsurface. These facilities are governed by the RCRA regulations. In 1987, the General Accounting Office indicated that nearly 5000 RCRA facilities had known chemical releases and that an estimated 50% of those facilities would require remediation to protect human health and the environment.[2] The RCRA program, like the Superfund program, expended many millions of dollars in the 1980s, but few sites were remediated.

For many corporations, the cost of remediating their contaminated sites for which they have been named as a potentially responsible party may very well be their largest unrecorded liability.[3] This can make contingent environmental liabilities the most significant unavailable piece of financial information for a corporation's existing or potential shareholders, creditors, employees, management, and other users of the corporation's financial information.[3] As a result, the U.S. Securities and Exchange Commission (SEC) and the Financial Accounting Standards Board have demanded responsible recognition and disclosure of environmental liabilities. The SEC has promised enforcement for noncompliance. In June of 1993, the SEC issued Staff Accounting Bulletin No. 92 (SAB 92), *Accounting and Disclosures Relating to Loss Contingencies*.[3] As corporations work to comply with SEC requirements, the extent of corporate liabilities for environmental issues is becoming clearer. In 1997, the *McKinsey Quarterly* reported that corporate environmental liabilities in the United States stood at $250 billion.[4]

Due to the widespread publicity that many of these sites received, the large amount of money that was spent, the amount of time that passed, and the few tangible results that were identified, the impression among Congress and the public in the early 1990s was that more time and money are spent on litigation and studies than on the actual cleanup.[5] For the 1980s and early 1990s, this impression may be true. The *New York Times* reported in its April 26, 1992 issue that of the $470 million that insurance companies processed in 1989 on Superfund-related claims, $410 million went to legal costs to defend their policyholders and dispute whether their policies covered the cleanup costs.[6]

In Chapter 2 of this book, the various factors that complicate environmental remediation projects and increase project costs are discussed in detail. However, as indicated above, the complicating issues are not all technical; they involve legal, political, and economic components as well.

1.2 THE ENVIRONMENTAL REMEDIATION INDUSTRY

Those companies that provide environmental consulting services, site assessments, remediation engineering, construction, and system operation are players in the environmental remediation industry, as are the various analytical laboratories, equipment manufactures, drilling companies, and a host of other service providers. The customers for these services include operators of waste disposal sites and owners of industrial manufacturing operations, oil refineries, chemical processing facilities, and commercial retail petroleum facilities, to name only a few. The various environmental laws and regulations that have been developed to ensure the protection of human health and the environment are drivers for this industry.

Since its beginnings, the environmental remediation industry has been complex, dynamic, rapidly evolving, and rich in technical challenge and uncertainty. In the early days, the challenges and uncertainties were considered so monumental that nearly all work was performed on a time and materials basis. Although costs were as much of a concern then as they are now, early on, most site owners and remediation service companies felt that there was little that could be done to control costs or accelerate project completion. In addition, most professionals involved in the early days of the industry felt that it was simply impossible to make reasonable estimates of the ultimate cost and time to complete a remediation project. In fact, many people still believe this to be true.

Fortunately, the industry has evolved a great deal since the late 1970s. A tremendous amount of innovation in site investigation and remediation technologies has occurred. In addition, the environmental laws and regulations which originally gave birth to the industry have evolved, in many cases becoming more pragmatic. However, despite technical successes and positive regulatory changes, the cost to achieve cleanup at any given site continues to remain high. Perhaps of even greater concern, environmental remediation projects have become notorious for their high budget overruns and long schedule delays. This is a problem not only for site owners and environmental service companies, but also the general public since ultimately it is the public that bears the increased costs of government and industry in the form of increased taxes and increased costs of goods and services. Benefits that accrue from reduced project cost and duration will be shared by all.[7]

1.3 THE CHANGING MARKET FOR REMEDIATION SERVICES

Some of the greatest changes in the environmental remediation industry have been in the structure of the market itself. The customers for environmental remediation services have become increasingly sophisticated regarding environmental issues and the services provided by remediation service companies. Many of these customers have become increasingly dissatisfied with high cost overruns, long schedule delays, and the apparent lack of direction commonly associated with remediation projects. In an attempt to control costs, these customers have adopted more stringent procurement techniques and types of contracts. Time and materials contracts are a thing of the past. Unit cost and firm fixed price contracts are becoming the rule, and many site owners are seeking environmental service firms that will commit to lump sum cost to completion. In general, customers are in search of straight answers to tough questions such as:

- How can we improve project performance?
- How can we ensure that we have the best overall strategy to achieve site cleanup?
- How much is it ultimately going to cost to clean up this site?
- When will this project actually be finished?

Those remediation service companies that can provide answers to these questions will achieve the greatest level of success in what has become a highly competitive marketplace.

A brief history indicates how rapidly the environmental remediation services market has changed. In the mid-1980s, the environmental remediation industry was gaining much momentum. At that time, the environmental service industry was the talk of Wall Street, as stock values for the entire sector continued to climb. A shortage of environmental professionals occurred as traditional engineering and construction companies made efforts to enter into what appeared to be a lucrative market. By the late 1980s, a full-scale boom was in swing; environmental professionals with only a few years of experience could command highly competitive salaries, with new job opportunities appearing all the time.

Unfortunately, the tremendous boom experienced by many remediation companies during the 1980s has been replaced by increased competition, reduced profits, and in some cases corporate downsizing. Most environmental service companies are in search of new ways to survive and prosper in the new market. This means that companies must learn how to realize a profit in the face of lower billing rates and fixed price contracts, even though the technical complexities and uncertainties associated with remediation projects have not significantly diminished.

The solutions that industrial/commercial site owners and remediation service companies (as well as the regulatory agencies and the public in general) are looking for can be found in modern project management techniques. Although there will continue to be advances in investigation and remediation technology, it is unlikely that a silver bullet "do-all" technology will be found. The area for significant cost savings will be found more in the management of technology implementation than in the technology itself. Therefore, industrial site owners will want to ensure that the environmental service companies they hire are skilled in modern project management processes.

In a similar fashion, although it is possible for a particular company to realize improved profits from the introduction of a new investigation or remediation technology, it unlikely that a sustainable business can be built on a particular technology innovation. This is not to discourage innovation but only to point out that as environmental service companies struggle to find new ways to survive and prosper in the new market, the answers they are looking for are to be found in improved project performance achieved by implementation of modern project management processes.

1.4   WHAT IS MODERN PROJECT MANAGEMENT?

The idea that project management is a distinct discipline, with its own body of knowledge, representing a profession in its own right, may be new to some, in particular to those whose training and background are geared to scientific understanding and technical development rather than coordinating and directing the activities of others to meet predefined goals or objectives. To such individuals, it may seem curious that the adjective "modern" is utilized since it implies that there is a more traditional form of project management, which has in some way been transformed and improved upon. This is indeed the case.

What is referred to as traditional project management had is beginnings in the defense and aerospace industries and came to be utilized heavily in the construction industry as well. Traditional project management is characterized by a focus on three main areas: cost, schedule, and achieving specifications. Modern project management, on the other hand, has a much broader range of focus, dealing with issues of quality, risk management, human resources, leadership, organizational structure, and information systems (to name only a few).

Project management as a distinct discipline and profession has been developing since some time in the late 1950s. Like the environmental remediation industry, it is relatively young when compared to disciplines such as geology, biology, engineering, accounting, and classical management. In the latter part of the 1990s, the issues of concerns for the profession are related to:

- Corporate implementation of modern project management processes and information systems
- Building and leading successful project teams
- Identifying and responding to project variances early in project execution
- Planning for and managing project risks

Perhaps the two areas where modern project management differs greatly from traditional project management are the definition of project success and the business responsibilities of the project manager. Project success is no longer defined merely as having completed a project within schedule and budget while meeting specifications. With modern project management, the customer's overall satisfaction with the outcome of a project is a significant if not the single most important success factor. Another success factor, which relates directly to the project manager's business responsibilities, is whether or not the project earned its expected profit and paved the way for new avenues of business growth. In other words, the project manager in modern project management has responsibility for customer satisfaction as well as business profit and loss.

1.5   NEED FOR INDUSTRY-SPECIFIC PROJECT MANAGEMENT APPLICATIONS

There are many books on project management theory as well as a host of articles on the subject of modern project management. Unfortunately, the books and articles currently in existence tend to deal with general project management theory or with major industries, such as construction and defense contracting. In addition, industry seminars and training programs tend to focus on generic project management practices. Many individuals find it difficult to go from the theory presented in these books and seminars to practical application in their specific industry or company. This problem is gaining recognition, and more individuals are studying and writing about the various nuances of applying project management theory in specific industries and organizational types.

1.6   THE ORIGINS OF THIS BOOK

This book grew out of a three-day training course on project management techniques for environmental remediation professionals developed by the author in 1994. The initial attendees of these courses were project managers working for environmental service companies. Later, this course was opened to environmental managers from larger industrial companies who were interested in both the techniques being presented and methods of evaluating the service companies they were hiring (i.e., did the remediation service companies have a firm grasp on modern project management processes and did they apply them in a formal, disciplined way).

Although a significant amount of project management theory was presented in the three-day course, at the request of the attendees, every effort was made to provide practical tools and techniques geared specifically to the needs of the industry. Therefore, whenever a project management theory was discussed, actual examples demonstrating application to the industry were provided. Every effort was made to write this book in keeping with this tradition.

As the training program evolved, and there was more interface with the companies receiving the training, it became apparent that training is not enough to ensure that project management will improve within a given company. What is needed is a systematic review and adjustment of overall business processes to ensure that support systems are in place to meet the needs of the project management group. In order to create these support systems, companies must define their project management processes, develop companywide project management information systems, and in many cases undergo organizational restructuring.

Finally, as the course further developed, it became clear that for the attendees to truly understand the improvements that must be made, it is helpful to review the history of the industry, some of the technical fundamentals, and to step back and think about the issues that have created so much challenge and technical uncertainty.

1.7 OBJECTIVE AND SCOPE OF THE BOOK

The overall objective of this book is to help improve the performance of environmental remediation projects through increased understanding and sound implementation of modern project management planning processes geared specifically to the environmental remediation industry. The benefits of improved project performance through the use of modern project management techniques will accrue to site owners in the form of reduced cost, shorter duration, increased project control, and improved communication. The benefits of improved project performance through modern project management will accrue to remediation service companies in the form of increased client satisfaction, improved profitability, increased employee morale, and improved projections of profitability and company backlog.

In terms of scope, it should be understood that the focus of this book is on modern project management theory, tools, and techniques and their practical application in the environmental remediation industry. One of the most important functions of the project manager is planning, and sound planning is the largest single factor for project success. Therefore, the emphasis of the theory, tools, and techniques is on *planning* rather than *execution*. However, project management techniques related to execution, such as the use of turnaround documents and the application of earned value analysis, are included, since they demonstrate the importance of sound project planning.

In order to be effective when communicating with project team members, project managers must have a working knowledge of the various technical details in their particular industry, although some speaking from a pure theory of management standpoint would disagree. Therefore, discussions of issues such as investigation techniques, remediation technologies, environmental regulations, and risk assessment processes are included in this book. However, these topics are discussed in a general way, in the context of project management, and with regard to the influence they have on the project manager's decision-making processes. In-depth details of such topics are left to publications that specialize in those areas.

1.8 FORMAT OF THE BOOK

The chapters of this book are arranged in a sequence that will have the greatest impact on understanding:

- The drivers, complexities, and nuances of the environmental remediation industry

- Project management theory
- Planning processes
- Planning under uncertainty
- Methods of implementing modern project management support systems

Chapter 2 presents the history and development of the environmental remediation industry in terms of regulatory drivers, technology developments, issues creating uncertainty, and current status of the industry. This is done to aid those who are new to the industry in understanding where the industry has come from and where it is likely to head in the coming years. Those who have been working in the industry for a long period of time may want to merely skim this chapter or proceed directly to Chapter 3. The last section of this chapter includes general guidelines for keeping project costs to a minimum. This section may be of interest to all, especially the customers of environmental remediation services.

Chapter 3 provides an overview of general project management theory and relates the theory to environmental remediation. This chapter may be viewed as a primer on project management. It was included as an aid for those unfamiliar with the basic tenets of modern project management theory. Long-time project managers may choose to skim this chapter, reading only the portions that relate the theory to the environmental remediation industry.

Chapters 4 through 9 are sequenced in accordance with a standardized planning process (presented at the end of Chapter 3) that can be applied to all types of remediation projects. This standardized process is based on the work of Harold Kezner, Ph.D., one of the leading authorities on modern project management theory and author of *Project Management; A Systems Approach to Planning, Scheduling, and Controlling*. These chapter are written from the viewpoint of a project manager working within an environmental service company who must create detailed project plans in order to service customer needs. However, all can benefit from these chapters, including environmental managers who purchase remediation services. This is because these chapters speak directly to the practical application of project management to environmental remediation. Each of these chapters contains techniques, guidelines, and rules of thumb for facilitating the overall process. Each of these chapters concludes with exercises which facilitate further development of the techniques presented. With the exception of Chapter 9, these exercises draw on the project data summary contained in Appendix A. The exercises in Chapter 9 pertain to earned value analysis. The necessary data for performing the exercises are contained therein.

Chapter 10 provides advanced project planning techniques for managing project risks and uncertainties. Included are techniques for comparing alternative project strategies and predicting project outcomes in terms of cost and time to project completion. A complete process for performing these evaluations and predictions is provided. The process is particularly applicable to brownfield redevelopment sites, real estate transactions, and bidding on lump sum cost to closure projects.

The techniques utilized by this process include systems modeling techniques such as decision tree analysis, Monte Carlo simulation, and influence diagramming as well as traditional project management techniques presented in Chapters 4 through 8. To aid in understanding the systems modeling techniques, discussions of strategic planning and systems thinking are provided, as are the fundamentals of probability theory. In order to demonstrate this process, this chapter finishes with a case study of its application to a real estate transaction.

Finally, Chapter 11 provides guidelines for implementing modern project management processes within an organization. These guidelines are provided by way of an actual case study implementation within a large environmental service company.

REFERENCES

1. Wallace, W.A., New paradigms for hazardous waste engineering, *Remediation,* Vol. 1, No. 4, 419, 1991.
2. General Accounting Office, Hazardous Waste, Corrective Action Clean-ups Will Take Years to Complete, GAO/RCED-88-48, U.S. General Accounting Office, Gaithersburgh, Maryland, December 1987.
3. Cummins, C.F. and Gladstone, D., The estimation and reporting of environmental contingencies, *Environmental Quarterly,* Vol. 8, No. 4, 5, 1996.
4. Merkl, A. and Robinson, H., Environmental risk management, *The McKinsey Quarterly,* No. 3, 152, 1997.
5. Wallace, W.A., New paradigms for hazardous waste engineering, *Remediation,* Vol. 1, No. 4, 420, 1991.
6. Little of Superfund settlements go to clean-up, *New York Times,* April 26, 1992.
7. Havranek, T.J., Hanish, M.B., and Fournier, L.B., Project management for cost effective environmental remediation, *Proceedings Project Management Institute 1992 Annual Seminar/Symposium,* Project Management Institute, Drexel Hill, PA, 1992, 120.

Strategic Issues in Environmental Remediation

2

This chapter is divided into four main sections. The first provides a discussion of significant regulatory programs that acted as the drivers for the industry and the rationale behind them. The second section discusses some of the fundamentals of environmental remediation, including site investigation, contaminant fate and transport, and remedial technology selection. In each of the first two sections, the advances that occurred as the industry progressed are discussed. The third section addresses the factors that create uncertainty and risk in cleaning up a contaminated site. The fourth section provides a number of strategic planning guidelines. These guidelines draw upon the fundamental concepts of environmental remediation and work to address the issues of uncertainty and risk.

This serves as a backdrop for understanding why project management is more important now than ever before and an introduction to the planning concepts and techniques that are built upon in Chapters 4 through 10.

2.1 ENVIRONMENTAL LEGISLATION AND REGULATIONS

To those involved in the fast-paced, complex, and ever-changing world of environmental remediation, it may seem hard to believe that the significant laws that served as the catalyst for the industry had yet to be written less than 25 years ago. The origins of the environmental remediation industry can be traced to the increased environmental awareness of the late 1960s and early 1970s, which led to a host of federal and state laws. This is not to say that environmental laws were not in place prior to the 1970s; for example, the Clean Water Act of 1972 can be traced back to the Rivers and Harbors Act 1899.[1] The environmental laws passed during the 1970s were more proactive and comprehensive in scope than their predecessors.[1] In addition, the EPA, created by President Nixon in 1970, would now act as a single entity to consolidate the various environmental programs.[1] The various federal and state laws are now integrated to such a degree that it can safely be said that there is never a time in the life of a chemical when it is not subjected to a some specific environmental regulation.[2]

The terms environmental law and regulation are often confused and used interchangeably; there is a distinct difference between the two. A law, or statute, is an act passed by Congress which provides the overall structure of a particular program and defines who has authority to administer it.[3] A regulation, on the other hand, is a formal requirement issued by the EPA or another federal or state agency (that has administrative authority over the program), to clarify the details of the law.[3]

2.1.1   Rationale for Environmental Legislation and Regulations

In order to understand the rationale for promulgating environmental legislation and regulations, we must focus on the role of the market economy, the problem of the commons, and the long-term viability of the environment.[3a] According to Jain:[4]

> Since labor and capital are scarce resources, their consumption is minimized by industry. Since the environment is, or rather has been in the past, an essentially free resource, its consumption has been ignored. Consequently, there has been considerable environmental degradation with attendant economic and social costs. Simply put, some economic and social costs are just not reflected in the market exchange of goods and services; also one can simply not ignore the third-party interests when looking at two party transactions of the buyer and the seller (existence of externalities). This is in fact the case with many environmental problems, and, thus, the transaction results in "market failure." Basically, market failure could result from high transaction costs, large uncertainty, high information costs, and the existence of externalities.[5] In order to correct market failure, two choices exit. One can try to isolate causes of the failure and restore, as nearly as possible, an efficient market process (process oriented) or alternatively bypass the market process entirely and promulgate regulations to achieve a certain degree of environmental protection (output oriented).
>
> Some environmental legislation and regulations are needed to protect the health and welfare of society, and market incentives will not

work. For example it would be very difficult to put a dollar value on discharge of toxic materials such as PCB or mercury to the environment. Another reason for environmental legislation and regulations is that long-term viability of the environment is essential in order to provide necessary life-support systems. Investment decision can rarely be made to take into account long-term protection of life-support systems which belong to everyone — a property of the commons.*

Whenever environmental issues are presented by the news media of the 1990s, areas of disagreement are often emphasized, rather than common threads and workable solutions. Regardless of political leanings, most would agree that they want a strong economy and low unemployment, along with a healthy and safe environment. The real question is how to achieve these goals. Since environmental legislation and regulations appear to be needed, it is important to find ways to structure them so that they have a minimal effect on industry yet still achieve the protective goals. The most common concerns and criticisms regarding environmental legislation and regulations include:[6]

- The costs outweigh benefits.
- Proper incentives for achieving social goals are not provided.
- Many programs involve unnecessary paperwork, creating additional costs.
- Programs are duplicated, or incompatible, at different levels of government, creating unnecessary costs and inefficiencies.

2.1.2 Evolution of Environmental Legislation and Regulations

In order to address concerns and criticisms such as those listed above, regulatory agencies are constantly modifying environmental regulations and new amendments to environmental legislation are continually being enacted. This constant change is one of the factors that creates uncertainty in environmental remediation projects. Fortunately, regulatory changes made during the 1990s have, for the most part, been more practical than those made during much of the 1970s and 1980s.

In the 1990s, there has been a concerted legislative effort in many of the states to move away from stringent cleanup requirements.[7] Many of these new legislative programs are designed to address so-called "brownfield sites." A brownfield site lies somewhere between a "significantly contaminated site" and

---

* From Jain, R.A., in *Standard Handbook of Environmental Engineering*, Corbitt, R.A., Ed., McGraw-Hill, New York, 1990, 2.2. Reproduced with permission of The McGraw-Hill Companies.

pristine virgin land (greenfield).[6] The U.S. General Accounting Office estimates that there are between 130,000 and 400,000 brownfield sites in this country.[7] What accounts for this discrepancy is that many of these sites may not be contaminated at all but merely perceived to require a prohibitively expensive cleanup before redevelopment can take place.[7] Essentially, many state legislators have questioned the efficiency of making brownfields so clean that "preschoolers can eat the dirt at recess," even though the site may be developed into a parking lot or for commercial use with limited possibility of dermal contact.[7] Prior to these new state programs, most developers and lending institutions were unwilling to take on the stringent municipal, state, and federal regulations required for remediation of brownfields.[7] This frustrated the goals of many communities seeking to redevelop abandoned urban property and at the same time caused new development to take place on greenfields rather than using areas with preexisting utilities and infrastructure.

The new less stringent state programs often provide incentives for site owners to perform voluntary cleanups and often make use of risk-based cleanup standards. Risk-based cleanup standards take into consideration planned redevelopment scenarios rather than simply deferring to background levels or maximum contaminant levels. A number of state programs now provide the owner with a written release from environmental liability as long as certain preestablished cleanup standards have been met. Such changes are altering the traditional drivers of environmental remediation projects. During the next decade, remediation projects may be driven more by the desire to redevelop active properties or restore dormant properties than merely by issues of regulatory compliance.

2.1.3 Significant Environmental Statutes

An in-depth presentation of environmental legislation and regulations is beyond the scope of this book. However, a brief overview of some of the more significant environmental legislation that led to the development of the industry is included, as it provides a context for later discussions on project management. Significant environmental statutes that pertain to environmental remediation include:

- The Clean Air Act of 1970 (CAA)
- The Federal Water Pollution Control Act of 1972 (now referred to as the Clean Water Act [CWA])
- The Safe Drinking Water Act of 1974 (SDWA)
- The Resource Conservation and Recovery Act of 1976 (RCRA)
- The Toxic Substances and Control Act of 1976 (TSCA)

■ The Comprehensive Environmental Response, Compensation, and Liability Act of 1982 (CERCLA)

The first three laws listed above (the CAA, CWA, and SDWA) are often referred to as end-of-pipe laws.[1] They were designed to protect the nation's air, surface water bodies, and water delivered for public consumption.

Although, the CAA, CWA, and SDWA represent great steps in the protection of human health and the environment, they are not in and of themselves the drivers of the environmental remediation industry. If they had been the only laws enacted in the 1970s, the environmental remediation industry, which focuses a great deal on restoring contaminated soil and groundwater, would not be as significant an industry as it is today.

As the CAA, CWA, and SDWA began to be implemented, a number of problems that were not addressed by these so called end-of-pipe laws were identified. These problems included:[2]

1. Many industrial facilities were typically paying others to remove and dispose of wastes that they could not safely release or dispose of on-site. In the early 1970s, a certain portion of the service companies involved in waste transportation, storage, and disposal were found to be not as sophisticated or fiscally sound as the commercial or manufacturing companies from which they were receiving materials. This led to a number of instances where wastes generated by conscientious customers were significantly mismanaged by unsophisticated and in some cases unethical service companies.

2. Many companies utilized chemical products or disposed of waste materials at a time when the environmental hazards associated with such materials were not well understood. As a result, these materials were handled, stored, and disposed of in ways which were later discovered to be detrimental to the environment.

3. Unplanned, accidental releases at a variety of industrial and commercial properties were creating new contaminated sites which needed to be addressed.

4. Contaminated sites were being discovered for which no owner or responsible party could be identified.

5. The EPA did not have a complete inventory of the hazardous materials that were being produced or utilized by various industries.

2.1.3.1   *The Clean Air Act of 1970*

The Clean Air Act of 1970, which amended the Air Quality Act of 1967, was established "to protect and enhance the quality of the Nation's air resources so as to promote public health and welfare and the productive capacity of its population."[8] This act has been amended a number of times since 1970, with amendments in 1974, 1977, and 1980.

In order to achieve the stated objective, the CAA required that the EPA establish the national level of air quality that must exist anywhere within the United States.[9] The EPA complied through the development of the primary and secondary National Air Ambient Quality Standards (NAAQS). The primary standards define levels necessary to protect human health, while the secondary standards define levels necessary to protect public welfare from any known or anticipated adverse effects of a pollutant.

The CAA delegated the responsibility for actually meeting and maintaining the standards to each state. The states were required to pass their own regulatory programs to achieve compliance with the NAAQS. These programs were to address issues of how to improve air quality in areas not meeting NAAQS and protect areas meeting the NAAQS from deterioration. Each state has its own set of regulations. Those involved in environmental remediation encounter these regulations when permitting air discharges from a soil or groundwater remediation system. These regulations can affect the choice of air treatment technology and the associated costs of installation and operation.

2.1.3.2 The Clean Water Act

The CWA is the primary authority for the nation's water pollution control programs.[10] The objective of these programs is to "restore and maintain the chemical, physical, and biological integrity of the Nation's waters."[9]

The development of effluent standards and permit systems is the key provision of the act.[11] The act set forth a program of establishing stream water quality standards and effluent limitations. The water quality standards are set by each state based on criteria that evolved from the Water Quality Act of 1965. The development of a water quality standard is a process of:[12]

1. Determining the designated water use
2. Adopting applicable criteria to maintain that use
3. Developing an implementation and enforcement plan for the adopted water criteria

Effluent discharge limitation guidelines and standards have been and are being developed to help meet stream standards and associated use designations. The enforcement mechanism legislated by the CWA of 1977 is the National Pollution Discharge Elimination System (NPDES). NPDES permits are issued to municipal and industrial dischargers to ensure that pollutant discharges do not result in a violation of water quality standards. Compliance with standards and permits is accomplished through state and federal monitoring, inspection, and enforcement programs. As with the CAA, the CWA emphasizes state primacy for permitting and enforcement. Those involved in environmental remediation projects often encounter the CWA when there is a need to discharge treated groundwater to a local stream.

2.1.3.3   The Safe Drinking Water Act

The SDWA was designed to protect drinking water supplies through the development of protective standards for public drinking water and groundwater. The drinking water standards are of two forms: the primary drinking water standards are health based and the secondary drinking water standards are aesthetically based. In general, the SDWA required the EPA to develop a list of contaminants to be regulated. The EPA was further required to develop *maximum contaminant level goals* (MCLGs) and *maximum contaminant levels* (MCLs) for each contaminant on this list. MCLGs represent goals that treatment systems should try to achieve; however, they are nonenforceable. MCLs are enforceable levels determined to be protective of human health. They are set as close as feasible to the MCLGs, but are based on best available technology. Finally, the SDWA provided directives for the EPA to establish regulations for the underground injection of fluids, both hazardous and nonhazardous, and for the protection of sole-source aquifers.

The SDWA was designed with the intent of the states having responsibility for enforcement. The states are required to submit drinking water programs to the EPA for approval. The states are also given the enforcement responsibility for protecting underground sources of water supply. This includes responsibility for permitting underground injection wells. Many of the states utilized MCLs as the basis for cleanup standards for remediation projects.

2.1.3.4   The Resource Conservation and Recovery Act of 1976

RCRA as it exists in the late 1990s is the culmination of a long series of legislation.[13] This legislation dates back to the Solid Waste Disposal Act of 1965. In addition to the act of 1976, other significant events in the development of the RCRA and its associated regulations include:

- The Hazardous Waste Regulations (1980)
- Used Oil Recycling Act (1980)
- Hazardous and Solid Waste Amendments (1984)

The key provisions of the RCRA include:

1. Hazardous wastes are clearly identified by definition and publication.[13] The regulations which identify whether or not a specific waste is hazardous can be found in Title 40 of the Code of Federal Regulations (40 CFR part 261.10). By definition, a waste is a hazardous waste if it meets the criteria of ignitability, reactivity, corrosivity, or toxicity defined in 40 CFR parts 261.21 through 261.24. In addition, a waste is considered a hazardous waste if it contains in any of the four lists published in 40 CFR parts 261.31 through 261.33. These lists include wastes from nonspecific sources, specific industrial processes, and pure or off-specification commercial chemical products.[13]
2. Hazardous wastes must be tracked from the time of generation through transportation, storage, and disposal.
3. The development of standards and regulations for hazardous waste treatment, storage, and disposal facilities, including facility design, groundwater monitoring, corrective actions, and financial responsibility for closure. These standards are codified in 40 CFR parts 264 and 265.
4. The creation of regulations for underground storage tanks, including leak detection; prevention of releases caused by spills, overfills, etc.; and corrective actions. These regulations are found in 40 CFR part 280.

RCRA in general deals with active sites where the owners and operators are easily identified. These regulations work toward correcting historical problems associated with unsophisticated and poor management practices at hazardous waste storage, treatment, and disposal facilities. The inclusion of the underground storage tank (UST) regulations created a mechanism for addressing releases due to degraded equipment. As such, RCRA has been a significant driver of the environmental remediation industry.

2.1.3.5   *The Toxic Substances Control Act of 1976*

TSCA provides a framework for developing an inventory of the types of chemicals produced and used by industry and for determining if newly produced chemical substances pose a potential threat to human health and the environment. Under this law, a chemical manufacturer must establish the safety of a given product and the EPA retains final authority to prohibit or severely restrict the use of a chemical. This law applies to commercial chemicals utilized by industry and not pharmaceuticals which are regulated by the Food and Drug Administration.

This law was enacted to prevent occurrences such as those associated with asbestos and polychlorinated biphenyls (PCBs) whereby these substances were heavily utilized by industry before it was known that they could present a significant hazard to human health and the environment. In addition to its requirements for chemical reporting and record keeping, TSCA provides for chemical-specific regulatory programs for asbestos, PCBs, and lead-based paint.

2.1.3.6 Comprehensive Environmental Response, Compensation, and Liability Act of 1980

CERCLA was designed to address the problem of inactive hazardous waste sites. While RCRA addresses those active sites where owners and operators are easily identified, CERCLA deals with those inactive sites where the responsible parties may be difficult to identify or may no longer exist. As might be expected, this program, like RCRA, has been a major driver of the environmental remediation industry. Environmental service companies are often called on to help those identified as responsible parties to perform the necessary response actions or to demonstrate that the EPA was in error in identifying them as a potentially responsible party.

2.2 FUNDAMENTALS OF ENVIRONMENTAL REMEDIATION

Remediation projects are often categorized by the legislative program under which they are being performed (i.e., RCRA, CERCLA, state UST, or voluntary cleanup program). The types of activities that must be performed are essentially the same for each type of project, yet the names of the various project stages and tasks change based on the legislative program. The names of the project stages and tasks associated with various legislative programs are discussed in Chapter 3 (see Section 3.9.1 and Table 3.2) and again in Chapter 6 (see Section 6.8.4 and the work breakdown structure templates provided in Appendix 4). Regardless of the name applied to the various project tasks, they are performed in order to:

1. Determine if any releases have occurred which require further investigation
2. Characterize the nature and extent of impact as well as the rate of contaminant migration
3. Evaluate cleanup alternatives, including cleanup standards, and select appropriate remediation technologies

4. Implement remedial actions, including system installation and operation (if necessary)

5. Document that intended goals have been achieved and discontinue remedial actions (project closure)

On the surface, it would appear that this would be a rather straightforward process. Individuals relatively new to environmental remediation might be able to guess the areas of potential project savings, such as:

- Perform only enough investigation to characterize the site physically and understand the nature and extent of environmental impact.
- Establish cleanup targets which are protective of human health and the environment; do not clean up any more than is necessary.
- Select the remediation technologies that are most cost efficient in achieving established cleanup goals.
- Implement remediation technologies in the most cost-efficient manner possible.

Unfortunately, industry experience indicates that this is much easier said than done due to a number of complicating factors. These complicating factors, which include technical, political, and economic components, are discussed in detail in Section 2.3. Before discussing these complicating factors, it is necessary to review some fundamental concepts regarding the fate and transport of environmental contaminants and the relative ease by which they can be removed or destroyed. In addition, a brief review of investigation techniques and remediation technologies is provided.

2.2.1 The Four Phases of Environmental Contamination

The variety of contaminants that can be released to the subsurface environment include organic compounds (volatile and semivolatile), inorganic compounds, and elements. Each contaminant has its own distinct set of physicochemical characteristics that govern its behavior in the environment.[14] In addition, each site has its own set of subsurface characteristics, hydrogeological and geochemical, which interact with the contaminant characteristics to ultimately determine contaminant distribution, fate, transport, and ease of removal/destruction.

An important fundamental concept in environmental remediation is contaminant partitioning. Contaminant partitioning refers to the separation of the total mass of released contaminants into the various phases in which they may exist in the environment. Before continuing, the words "phase" and "component" should be defined. For our purposes, a phase can be defined as any physically distinct part of a system that is separated from other parts of the system by definite bounding surfaces.[15] Therefore, gasoline floating on the top groundwater can be referred as a separate liquid phase. The word component refers to the individual constituents which sum together to compose a phase. Referring again to gasoline, although a separate liquid phase may be observed, this phase is made up of a large number or organic components.

When organic compounds are released to the environment, they will partition themselves into four distinct phases:

- Vapor phase, within the nonliquid-filled pore spaces of the soil or as distinct bubbles within the liquid-filled pore spaces
- Adsorbed phase, adhered to the soil particles
- Free product, or nonaqueous-phase liquid (NAPL), not adhered to soil and as a distinct separate phase
- Dissolved phase, within the groundwater

Figure 2.1 presents the four phases of impact as commonly depicted for a material such a gasoline which has a density less than that of water and consists of components that are volatile enough to enter the vapor phase and soluble

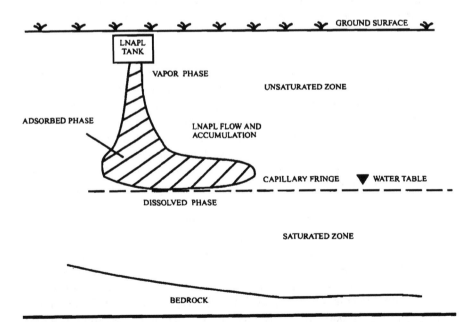

FIGURE 2.1 Behavior of LNAPL in the subsurface environment. (From Sutherson, S.S., *Remediation Engineering: Design Concepts,* Lewis Publishers, Boca Raton, FL, 1997, 17. With permission.)

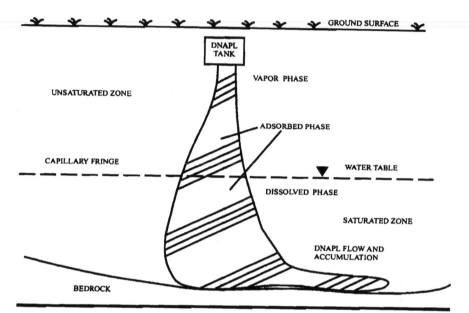

FIGURE 2.2   Behavior of DNAPL in the subsurface environment. (From Sutherson, S.S., *Remediation Engineering: Design Concepts,* Lewis Publishers, Boca Raton, FL, 1997, 17. With permission.)

enough to dissolve into the water. The free or separate mobile phase is often referred to in the industry by the acronym NAPL (rhymes with apple). This term is said to have been originally coined during studies of a hazardous waste land-fill in New York.[16] Organic compounds such as benzene or product mixtures such as gasoline which are "lighter," meaning less dense, than water are referred to as LNAPLs, while chlorinated solvents such a trichloroethylene (TCE) which have a density greater than water are commonly referred to as DNAPLs. Figure 2.2 presents the four phases of impact commonly depicted for a material such as TCE.

The various equations for contaminant partitioning, which are a function of contaminant as well as subsurface properties, predict, and industry experience verifies, that the percentages of the total mass of released materials do not partition equally into four different phases. An understanding of the mass of the contaminants within each phase is an important consideration in developing cost-effective remediation strategies. Table 2.1 represents the estimated phase distribution of a 30,000-gallon gasoline spill in a medium sand aquifer, with a depth to the water table of 15 ft^2.

A review of this table indicates that remedial actions focused on the free product and adsorbed phases will address 95% of the released gasoline, while

TABLE 2.1   Phase Distribution of a 30,000-Gallon Gasoline Spill

Phase	Contaminated spatial volume (yd³)	Percent of total	Contaminated gasoline volume (gal)	Percent of total
Free product	7,100	1.0	18,500	62
Adsorbed (soil)	250,000	20.0	10,000	33
Dissolved (water)	960,000	79.0	333	1–5
Vapor	Not quantified	—	—	<1

From Sutherson, S.S., *Remediation Engineering: Design Concepts,* Lewis Publishers, Boca Raton, FL, 1997, 4. With permission.

those focused on groundwater will address less than 5%. This table also provides data on the spatial volume of the various phases. Note that the dissolved phase represents 79% of the contaminated spatial volume although it contains less than 5% of the total mass of the released contaminant.

Although the examples provided in Figures 2.1 and 2.2 are for organic compounds, the concept of four-phase partitioning can also be applied to inorganic contaminants, depending on their physical properties. For example, solubility will determine the degree to which inorganic contaminants will dissolve in groundwater (as it will for organic compounds). Under the proper conditions of pH, concentration, etc., such contaminants could precipitate as a separate, albeit solid, phase. A vapor phase would be relatively rare; however, a substance such as elemental mercury can exist in the vapor phase. Finally, inorganic contaminants can adhere or adsorb to soil particles.

2.2.2   Moisture Zones in the Hydrogeologic System

The various phases described above exist within the soil and in particular within the pore spaces that exist between the soil particles. The total unit volume of a soil or rock consists of the volume of the solid portion (the soil particles) and the volume of the pore spaces (voids). Porosity is defined as the ratio of the pore volume to the total unit volume of the soil and is reported as a decimal fraction or percent. Figures 2.1 and 2.2 call attention to the fact that within the subsurface hydrogeologic system there exist three distinct subsurface zones which are characterized by the amount of moisture present in the pore spaces. The three zones are known as the saturated zone, capillary fringe (also referred to as the tension-saturated zone[17]), and unsaturated zone. In the saturated zone, the pore spaces are completely filled with water and the water saturation equals the porosity. This zone exists below the water table (i.e., the water level that would be measured by gauging a monitoring well). The unsaturated zone occurs above the water table, more accurately down to the capillary fringe. In this zone, the moisture content is less than the porosity, which means that the pore spaces are not completely filled.

The capillary fringe exists between the unsaturated zone and the saturated zone. This zone is due to capillary action. Capillary action is caused by adhesive forces between the groundwater and the surfaces of the soil particles which are greater than (i.e., dominate) the cohesive forces between the water molecules and themselves. The adhesive forces cause the water to attach itself and climb the vertical surfaces of the soil pore spaces. It can be said that the water "reaches up and tries to wet as much of the interior surface as it can."[18] In doing so, it rises above the general water surface level (the water table). The pore spaces within the capillary fringe are completely filled with water. The thickness of the capillary fringe varies depending on the type of soil present. In general, fine-grained soils such as clays have thicker capillary fringes than do coarser grained soils such as sand.

An important point that should not be overlooked with regard to the four phases of contaminant impact and the distinct groundwater zones is that although certain phases (dissolved, vapor, etc.) may be more commonly associated with a given subsurface zone (saturated, unsaturated), this does not mean that these phase are exclusive to a given zone. For example dissolved-phase hydrocarbons can be contained in the moisture present in the unsaturated zone. In a similar fashion, adsorbed-phase hydrocarbons may well exist in the saturated zone as well as the unsaturated zone. Figure 2.3 is included to demonstrate this point.

2.2.3   Factors That Affect Contaminant Distribution and Migration

The factors that affect contaminant distribution and migration can be grouped into three separate categories: contaminant properties, the properties of earth materials, and site characteristics.

2.2.3.1   *Contaminant Properties*

The following properties have a significant impact on how contaminants partition into the four phases, as well as their fate and transport in and recovery from the subsurface environment:

- Density
- Water solubility

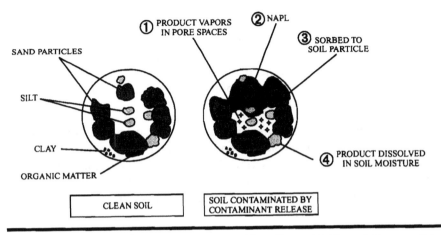

FIGURE 2.3 Phases of contaminants present in soil matrix. (From Sutherson, S.S., *Remediation Engineering: Design Concepts,* Lewis Publishers, (Boca Raton, FL, 1997, 34. With permission.)

- Vapor pressure
- Octanol–water partitioning coefficient
- Henry's law constant
- Biological oxygen demand
- Viscosity

Density. The density of a fluid is its mass per unit volume. In the centimeter-gram–second (cgs) system of units, density is measured in grams per cubic centimeter (g/cm^3) for liquids and grams per liter (g/l) for gases (vapors). In the absolute English system, the density of both liquids and gases is measured in pound mass per cubic foot (lbm/ft^3). For the purposes of environmental remediation, the reason for knowing the density of a liquid substance is to determine whether it will float or sink in groundwater (LNAPL or DNAPL) or whether a particular gas is heavier or lighter than air.[19] The density of pure water at 60°F is 1 g/cm^3, equivalent to 62.4 lbm/ft^3. The density of air at 70°F and 1 atmosphere is 1.20 g/l, equivalent to 0.075 lbm/ft^3. Various refined petroleum products range in density from 0.75 g/cm^3 (46.8 lbm/ft^3) for automotive gasoline to 0.97 g/cm^3 (60.5 lbm/ft^3) for No. 6 fuel oil.[18] Chlorinated solvents have much higher densities; examples include chlorobenzene at 1.1 g/cm^3 (68.62 lbm/ft^3) and tetrachloroethylene at 1.63 g/cm (101.66 lbm/ft^3)[3].

Aqueous solubility. Aqueous solubility determines the degree to which a chemical will dissolve in water (i.e., enter into the dissolved phase). Solubility, measured in milligrams per liter (mg/l), indicates maximum possible concentration. Above this concentration, two phases exist in the solvent (groundwater)–solute (chemical contaminant) system. In other words, above this concentration, a separate free phase will exist. High solubility indicates a low tendency to adsorb to soil. Low solubility is on the order of 1000 mg/l (e.g., benzene = 1800 mg/l) and high solubility is on the order of tens of thousands of milligrams per liter (e.g., methylene chloride = 13,000 mg/l).

Vapor pressure. Vapor pressure is a parameter that can be used to estimate the tendency of a compound to volatilize and enter into the vapor phase. It is measured in millimeters of mercury (mmHg). Low-volatility compounds have a vapor pressure less than 1 mmHg (e.g., benzo(*a*)pyrene = 5.60×10^{-9} mmHg), while high-volatility compounds have a vapor pressure greater than 5 mmHg (e.g., *m*-xylene = 10 mmHg).

Octanol–water partitioning coefficient. The octanol–water partitioning coefficient, measured in milliliters per gram (ml/g), indicates the tendency of a chemical to partition between groundwater and soil. A large octanol–water partitioning coefficient signifies a highly hydrophobic compound, which indicates strong sorption (e.g., PCB Arochor 1254 = 4.25×10^{4} ml/g).

Henry's law constant. Henry's law constant is somewhat analogous to vapor pressure. Whereas vapor pressure is a measure of chemical partitioning between a pure substance and its vapor phase, Henry's law constant is a measure of the tendency of a compound in solution to exit the solution and enter the vapor phase. In addition to having a bearing on contaminant partitioning, this parameter is used extensively in the design of air-stripping systems and the evaluation of soil vapor extraction systems.

Biological oxygen demand. Biological oxygen demand (BOD) can be used as an indicator of the relative biodegradability of a compound. It is measured in milligrams per liter (mg/l). BOD is a measurement of the dissolved oxygen used by microorganisms in the biochemical oxidation of organic matter. Compounds with a BOD less than 0.01 mg/l are considered recalcitrant, while those that have a value closer to 0.1 mg/l are considered biodegradable.

Viscosity. The viscosity of a fluid is a measure of its resistance to flow when acted upon by an external force such as a pressure differential or gravity. Viscosity is commonly measured in centipoise (cP). The viscosity of water at standard temperature and pressure is 1 cP. Fluids such as motor oil are very viscous (>100 cP), while automotive gasoline and jet fuel have low viscosities (<1 cP). Viscosity has a bearing on the movement of NAPL in the unsaturated zone, the spreading of a separate phase LNAPL or DNAPL plume, the flow of LNAPL into a recovery well, and the relative ease with which it can be pumped to the surface.

2.2.3.2   Properties of Earth Materials

Subsurface earth materials can be placed into two broad classifications: unconsolidated materials and consolidated bedrock. In general, unconsolidated material refers to the loose soil materials which result from the erosion of bedrock. These loose materials include gravel, sand, silt, and clay. Previously excavated materials and construction debris can also be included in the category of unconsolidated materials. Consolidated bedrock includes (1) sedimentary rocks such as shale, sandstone, and limestone; (2) igneous rocks such as granite; and (3) metamorphic rocks such as slate and marble.

The significant fluid-transmitting and contaminant-partitioning properties of the earth materials include:

- Porosity
- Hydraulic conductivity
- Organic carbon content

Porosity.   As previously stated, porosity is the ratio of the volume of void space in earth materials to the total unit volume of the material. It is typically expressed as a percentage or decimal percent. It is based on the grain size and shape of the earth materials. Table 2.2 indicates ranges of porosity values for different earth materials.

Porosity is strongly affected by the degree of material sorting and the shape of the particle grains. Sand that is well rounded and sorted such that all grains are nearly the same size will have a porosity on the higher end of the range provided in Table 2.2. On the other hand, if the sand grains are highly angular and interspersed with silt and clay particles, the porosity will be significantly reduced.

Hydraulic conductivity.   Hydraulic conductivity is a measure of the ability of a geologic material to transmit a fluid, dependent on the type of fluid passing through the material. Its value is influenced by the degree of interconnectedness of the pore spaces in unconsolidated materials or the crevices and fractures of consolidated materials, as well as the viscosity and density of the fluids flowing through the material. The hydraulic conductivity indicates the quantity of water (or other fluid) that will flow through a unit cross-section area of a porous medium per unit time and under a hydraulic gradient of 1 (hydraulic gradient is

TABLE 2.2   Range of Values of Porosity

	n (%)
Unconsolidated deposits	
Gravel	25–40
Sand	25–50
Silt	35–50
Clay	40–70
Rocks	
Fractured basalt	5–50
Karst limestone	5–50
Sandstone	5–30
Limestone, dolomite	0–20
Shale	0–10
Fractured crystalline rock	0–10
Dense crystalline rock	0–5

From Freeze, R.A. and Cherry, J.A., *Groundwater*, Prentice-Hall, Englewood Cliffs, NJ, 1979, 37. Reprinted by permission of Prentice-Hall, Inc., Upper Saddle River, NJ.

discussed in the section on site conditions). Equation 2.1 defines hydraulic conductivity in terms of permeability (k), density (ρ), acceleration due to gravity (g), and viscosity (μ):[20]

$$K = \frac{k\rho g}{\mu} \tag{2.1}$$

Note that permeability (k) is a function of the porous medium only and has the primary dimension of length squared (L^2). This parameter, which is also referred to as specific or intrinsic permeability, is widely used in the petroleum industry where the existence of gas, crude oil, and water in a multiphase flow system makes the use of a fluid-free conductance parameter attractive.[21]

Hydraulic conductivity in the cgs system of units is expressed in centimeters per second (cm/s), which is easily converted to meters per day. In the water well industry, units of gallons per day per square foot (gpd/ft^2) are commonly used. Figure 2.4 indicates the range of hydraulic conductivity for different media.

Organic carbon content. The organic carbon content of soil is the percentage fraction of naturally occurring organic carbon in soil. The higher this value, the greater the degree to which an organic compound will adsorb to the soil. The organic carbon content along with the octanol–water partitioning coefficient of

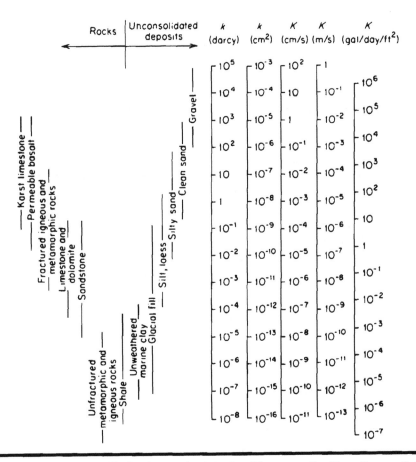

FIGURE 2.4   Range of hydraulic conductivity for different media. (From Freeze, R.A. and Cherry, J.A., *Groundwater*, Prentice-Hall, Englewood Cliffs, NJ, 1979, 29. Reprinted by permission of Prentice-Hall, Inc., Upper Saddle River, NJ.)

the contaminant of concern can be used to determine a third value: the soil-partitioning coefficient, which ultimately determines the degree to which an organic compound will be found in the adsorbed phase.

2.2.3.3   Site Characteristics

The site characteristics that have a significant impact on the fate and transport of contaminants in the environment as well as their partitioning into the four phases include:

- Depth to groundwater
- Hydraulic gradient
- Volume and rate of contaminant release

Depth to groundwater. In general, the greater the depth to groundwater, the greater the mass of an organic contaminant that will partition into the adsorbed phase (assuming that the contaminant in question has some propensity to partition into the adsorbed phase). Turning our focus to flow by the force of gravity alone (i.e., not considering dissolved-phase flow), the leading edge of the contaminant plume will move into the pore spaces of the unsaturated zone. The contaminant must exceed the adsorptive capacity of the pore spaces it has encountered before it will move further downward. The contaminant will continue moving in this fashion, leaving a trail of residual hydrocarbons. If the depth to water is great enough (or the volume of loss is small enough), a separate phase layer may not occur. In general, the greater the depth to groundwater, the lower the dissolved-phase concentrations in the groundwater.

Hydraulic gradient. The hydraulic gradient (I) is the difference in hydraulic head (groundwater elevation), $h_1 - h_2$, divided by the distance along the flow path:

$$I = \frac{(h_1 - h_2)}{L} \tag{2.2}$$

The hydraulic gradient, along with hydraulic conductivity and porosity, are utilized, by way of Darcy's equation, to determine the average linear groundwater velocity beneath the site. Darcy's equation is provided below:

$$V = \frac{KI}{7.5n} \tag{2.3}$$

where V = velocity (ft/day), K = hydraulic conductivity (gpd/ft^2), i = hydraulic gradient (ft/ft), and n = porosity (decimal percent). From this equation, it can be seen that groundwater velocity is directly proportional to the hydraulic gradient (i.e., the greater the hydraulic gradient, the greater the groundwater velocity).

Volume and rate of contaminant loss. The volume or total mass of the contaminant loss is a function of site operations and release scenarios. In general, a rapid loss of a large amount of material will cause accelerated downward migration and increased chance of separate phase product. A slow leak over time tends to spread more horizontally over the unsaturated zone, creating less chance of separate phase product.

2.2.4 Migration Processes

From the information provided in the previous sections, one can imagine that the contaminant properties and site characteristics interact in a multitude of complex ways, all of which have an effect on the migration rates of the liquid, vapor, and dissolved phases. A complete analysis of migration processes would be quite lengthy and beyond the scope of this book. Therefore, only a conceptual review is provided here.

Depending on the properties of the contaminant and the adsorptive capacity of the subsurface soil, a certain percentage of the released mass of the contaminant will exist as residual liquids in the unsaturated and saturated zones. This percentage of the mass will not migrate further unless something is done to disrupt adsorption, such as increasing the temperature of the subsurface environment. The free mobile liquid hydrocarbons migrate either at the top of the capillary fringe or along the impermeable base of the aquifer, as a function of density, viscosity, porosity, permeability, and hydraulic gradient. Vapor-phase migration will occur in the void spaces of the unsaturated zone primarily by way of gaseous diffusion, along a concentration gradient until an equilibrium is established. Finally, dissolved-phase migration will occur in both the unsaturated (infiltration of rain water) and saturated zone. In the saturated zone, migration will occur both as advection (movement with groundwater) and dissolution (movement along a concentration gradient). Soil adsorption will work to retard this process and natural biodegradation will work to limit the amount of material available for transport. Lastly, simple dilution processes will work to reduce the concentration gradient moving away from the contaminant source.

2.2.5 Environmental Assessment, Techniques, and Advancements

The primary objectives of any environmental investigation are to:

- Determine if a release has occurred
- Define the nature (type) and extent of contamination (adsorbed, free liquid, dissolved, vapor phases)
- Understand how the various phases are affected by hydrogeologic conditions
- Provide data necessary for the evaluation of remedial alternatives

Some of the most useful information from an environmental assessment comes in the form of maps, including:

- Distribution maps, showing exact sample results at different locations
- Iso-concentration maps, indicating contour lines of equivalent concentration
- Groundwater gradient maps, presenting lines of equivalent groundwater elevation
- Geologic cross-sections, depicting subsurface geology

There are a great variety of techniques to determine the analytic and physical characteristics of a site; these include:

- Soil sampling (surface and at depth)
- Groundwater sampling (via monitoring wells)
- Groundwater gauging (via monitoring wells, providing depth to water)
- Soil gas surveying
- Geoprobe surveying
- Slug testing (providing aquifer parameters, in particular hydraulic conductivity)
- Aquifer pump testing (providing hydraulic parameters of aquifer, including response to pumping)
- Surface geophysical surveying
- Subsurface (downhole) geophysical surveying (logging)

In the early days of the industry, the primary way of characterizing the site analytically and physically was by direct sampling of soil and groundwater and analysis in an off-site laboratory. Drilling rigs were utilized to collect soil samples at depth, often as part of a monitoring well installation procedure. The permanently installed monitoring wells were utilized for later groundwater gauging, sampling, and aquifer pumping tests. Although these techniques will always be part of an environmental investigation, soil gas surveys, geoprobe surveys, and on-site analysis represent cost-saving techniques that both accelerate and assist in providing a more thorough site investigation.

A soil gas survey is a screening technique that can be utilized to determine the horizontal extent of impact. The sampling techniques can be performed by two methods. The first involves inserting a probe, such as a stainless-steel tube, into the ground and withdrawing vapors with a vacuum collection device which discharges to a field meter such as a photoionization detector, a flame ionization detector, or even a field gas chromatograph for compound-specific analysis. This method is quick, relatively inexpensive, and (if performed properly) provides a real-time indication of the horizontal extent of impact. The results from the soil gas survey can be placed on a map and contoured (iso-concentration map). This map can be used to select locations for monitoring wells and confirmation soil sampling. Savings are incurred through the collection of fewer soil samples and better placement of monitoring wells. The second method of performing a soil gas survey involves the installation of buried accumulation devices, which are retrieved after a week or more and sent to an analytical laboratory. This second method is more accurate, but also more costly.

A geoprobe survey permits the rapid collection of soil samples at depth and the collection of groundwater samples without the installation of permanent monitoring wells. Its development marked a significant advance in cost-effective site investigations. The geoprobe is hydraulically powered and uses static force and percussion to advance sampling tools to depth. Samples are recovered for on-site screening and off-site laboratory analysis.

Slug tests and aquifer pumping tests provide aquifer parameters. The slug test is a low-cost method of determining hydraulic conductivity. Aquifer pumping tests provide additional aquifer parameters needed for the design of groundwater recovery systems.

Surface geophysics is commonly used to identify the location of an underground storage tank, piping, etc., as well as changes in lithology. Recent advances in surface geophysics allow the identification of areas highly saturated with organic compounds (separate phase layers).

2.2.6 Remediation Processes and Advancements

A large number of remediation processes currently exit for the treatment of contaminated environmental media. In an appendix to the book *Remediation Engineering: Design Concepts* (1997, Lewis Publishers), Suthan S. Sutherson provides a list of over 100 existing processes. In addition, new remediation technologies are constantly emerging. With so many options available, most sites will require a screening process and feasibility study to ensure that the most cost-effective process combinations are selected.

Although there are many possible remedial technology options available, in reality there are only four possible methods of dealing with contaminants in the environment:

- Removal
- Extraction
- Destruction
- Isolation

Removal refers to the physical removal of the contaminant along with the medium in which it resides. Examples of removal include contaminated soil excavation and groundwater pumping from a recovery well. Extraction refers to the direct removal of the contaminant from the environmental medium using techniques such as soil vapor extraction and soil washing. Destructive processes include technologies such as bioremediation and incineration. Isolation includes techniques to immobilize the contaminant, such as chemical fixation or the use of retainer walls.

The methods for dealing with contaminants in the environment are accomplished with remedial processes; all remedial processes can be placed in one of four categories:

- Physical
- Chemical
- Biological
- Thermal

For example, isolation can be accomplished by pozzolonic fixation (a chemical process) or *in situ* vitrification (a thermal process). Some remediation technologies may involve processes that fit into more than one category. Soil vapor extraction, for example, which can remove an adsorbed-phase volatile organic compound such as xylene (a physical process), will increase the concentration of oxygen in the subsurface environment and may enhance biodegradation of a contaminant as well (a biological process).

The most basic factors which determine technology selection include:

- Chemical constituents
  □ Volatile organic compounds
  □ Semivolatile organic compounds
  □ Metal
- Media type
  □ Soil
  □ Groundwater
  □ Vapor
- Media location
  □ Aboveground
  □ *In situ*
- Partitioning phase of interest
  □ Adsorbed
  □ Free mobile liquids
  □ Dissolved phase
  □ Vapor

From a practical point of view, there is no single remediation technology that will address all phases of contamination at a given site or within a given medium. An effective remediation strategy typically requires a combination of technologies and processes which take advantage of the properties of the contaminant, the subsurface environment, and their interactions which influence contaminant distribution, fate, and transport (migration). For example, factors that lead to high mobility, such as high vapor pressure, high solubility, and highly conductive porous formation, are the same factors that can be taken advantage of to enhance contaminant removal or extraction. On the other hand, factors that limit mobility, such as a highly viscous contaminant with a high octanol–water partitioning coefficient in a clayey formation, can be taken advantage of, perhaps through the application of isolation processes such as chemical stabilization.

Some of the greatest advances in remediation technology and remediation strategies have come from experiential understanding of the mass distribution and the fate of contaminants. An example is a large rapid release from an underground storage tank of a compound such as xylene into a subsurface environment consisting of sand, silts, and clays. Xylene is highly volatile, of low solubility, biodegradable, and of low density (LNAPL). In the late 1970s and early 1980s, the response might have been to install a groundwater pump and treat system along with pumps to collect free product floating on the surface of the water table. However, a significant mass of the contaminant would remain as an adsorbed phase in the unsaturated soil. This material could act as a continual source of dissolved-phase impact. However, at this point in time, many sight owners might be reluctant to excavate the material due to high off-site disposal costs. As the industry developed, it became apparent that *in situ* soil vapor extraction could be utilized to extract the adsorbed-phase xylene from the unsaturated soil. In addition, the increase in oxygen would encourage biodegradation of the contaminant. By the mid to late 1980s, when soil vapor extraction systems were combined with groundwater pumping/treatment and free product recovery systems for compounds or product mixtures of the same class as xylene, the cost and time to achieve site closure began to decrease significantly.

Air sparging was an emergent technology in the latter part of the 1980s. With this technology, it became possible to treat groundwater *in situ*. This technology is particularly applicable for biodegradable volatile compounds with a moderate to a high Henry's constant. Air sparging involves the injection of air into the saturated zone. The air bubbles move up through the saturated zone and into the saturated zone, where they are captured by a soil vapor extraction system. The contaminants are actually transferred from the dissolved phase and into the vapor phase. In addition, the introduction of oxygen present in the air encourages *in situ* biodegradation in both the saturated and unsaturated zones. Air sparging saves money in several ways. The first is that a treatment system is not needed to recover groundwater. The second is that is takes less energy to move air than it does water. The third is that the lack of a groundwater treatment system meant less equipment and lower operation costs. However, air-sparging systems are not applicable to complex subsurface geology. Subsurface barriers or zones of decreasing permeability could cause injected air to move in unexpected and undesirable ways, increasing the mobility of contaminants away from the site.

Continuing with our example, high-vacuum dual-phase extraction was developed by the early 1990s. With this technology, a high vacuum, much higher than standard soil vapor extraction, can be introduced into the subsurface. This high vacuum can physically remove groundwater, separate phase hydrocarbons and vapor-phase hydrocarbons, and at the same time introduce oxygen in both the saturated and unsaturated zone to encourage biodegradation. Although the technique is highly effective, it costs more the air sparging/soil vapor extraction, has more equipment that must be maintained, and involves surface groundwater treatment.

These three examples of remediation technologies demonstrate not only advances over time for a particular class of contaminants, but also demonstrate how an understanding of contaminate properties and subsurface migration processes impacts remedial technology selection.

2.3 COMPLICATING FACTORS IN ENVIRONMENTAL REMEDIATION

The remainder of this chapter draws heavily on a paper entitled "Project Management for Cost Effective Environmental Remediation" by Timothy J. Havranek, Louis B. Fournier, and Mark B. Hanish, originally published in the *Proceedings of the 1992 Seminar/Symposium of the Project Management Institute* (reprinted here, with modification, by permission of the Project Management Institute).

A variety of factors make it difficult to achieve cost-effective environmental remediation; these include:

- Differing objectives of the project stakeholders
- Charged political nature of these projects (due primarily to the high stakes involved)
- Conflicting regulations
- Tremendous uncertainty in the subsurface environment
- Unrealistic specifications
- Legal issues

2.3.1 Differing Objectives of Project Stakeholders

There is seldom one stakeholder involved in the cleanup of a hazardous waste site. In fact, it is more accurate to say that there is never a single stakeholder involved in a project. If public concern is high in regard to a particular site, there can be literally hundreds of stakeholders. A typical listing of stakeholders involved with most projects would include:

- Facility owner(s)/operator(s)
- Regulatory agency(s)
- Legal counsel
- The public
- Technical consultants

2.3.1.1 Facility Owner(s)/Operator(s)

It is common for industrial sites to have many past owners and operators. It is these individuals who, for the most part, will fund the actual investigation and cleanup. For this reason, they are the primary stakeholders. The facility owners/operators are in the business of making a profit by providing a useful product or service. Although most companies want a positive public image and to be seen as good corporate citizens, the reality is that money spent on environmental issues directly impacts overall profitability. Therefore, the objective of the facility owners/operators is usually to keep costs to a minimum, and rightly so. Owners/operators commonly indicate that they want to do the minimum required to satisfy the regulatory agency. Unfortunately, without a proper understanding of the issues at hand, doing only what is necessary to satisfy the regulators may require doing much more than is required to protect human health and the environment. If the environmental problem is at a site that has had multiple owners/operators/users (liable parties), each party usually has an additional objective of ensuring that it pays only for its contribution to the problem. When this is the case, the complexity can increase since each party may choose to separately hire legal counsel and technical consultants to assist in determining the appropriate contribution.

2.3.1.2 Regulatory Agency(s)

It is safe to say that the objective of the regulatory agency(s) is to maintain proper adherence to environmental regulations. Although most agency literature indicates that cost and technical feasibility are considerations, the agency for the most part must insist on strict adherence to the requirements of these regulations.

This desire for strict adherence stems from the fact that regulatory agencies are answerable to the public and must be able to defend decisions they make and activities they permit.

2.3.1.3 Legal Counsel

The facility owner/operator utilizes in-house or external legal counsel to assist in protecting its overall liability. In many cases, legal counsel may wish to avoid a commitment to clean up a site because hazards and costs may not be well understood and any cleanup activity may be construed by others as an acknowledgment of a severe hazard even if one has not been identified. Such commitments could inadvertently strengthen the case of any lawsuit that is brought against the facility owner/operator.

Although the above may sound like a questionable practice, that is not the case. At many of these sites, it can be extremely difficult to scientifically prove that a risk to human health actually exists. For example, now that Missouri's Times Beach site has been cleaned up, serious questions regarding the toxicity of dioxin, the contaminant of concern, have arisen. Scientific studies performed on dioxin since the site was cleaned up indicate that the substance is not as toxic as once believed.

2.3.1.4 The Public

The public's most common goal for a hazardous waste site is cleanup to pristine conditions. In general, the public sees any amount of environmental contamination as unacceptable. In addition, individuals often want to be reimbursed for perceived harms. This can result in a multitude of toxic tort cases.

2.3.1.5 Technical Consultants

The regulations that have been established for the cleanup of hazardous waste sites have resulted in the development of an entire industry which was virtually nonexistent less than 25 years ago. A variety of environmental consulting/ remediation firms have appeared on the scene over this time period. The objectives and approaches of these companies depend on their philosophy and background. Environmental projects are multidisciplinary and require a multidisciplinary approach. If an environmental consulting firm has access to appropriate technical specialists and is skilled in coordinating them, it will likely be in the best position of all the stakeholders to understand the nature of the environmental problem and the methodologies appropriate to address site cleanup. On the other hand, if an environmental consulting firm is overspecialized in a certain discipline, this firm may bias its environmental solutions toward the discipline with which it is most familiar. For example, a firm with a background in civil engineering may offer solutions that involve excavation, hauling, and disposal of contaminated material in a secure landfill. Another firm with a background in chemical engineering may offer solutions that involve excavation and treatment of affected environmental media with a mobile chemical treatment unit. Yet another firm, which employs numerous Ph.D.-level scientists, may study the site to an excessive degree before identifying appropriate solutions.

With this variety of stakeholders and their different points of view, it may be difficult to even begin a project because the stakeholders cannot agree on a common objective. It is often extremely difficult to define exactly what would or should constitute success in an environmental project, since each stakeholder would define success differently.

2.3.2 Charged Political Nature of Environment Remediation Projects

The politics of an environmental project can be quite complex. Media and public attention focused on an environmental problem may cause a local or regional public official to make the cleanup of a hazardous waste site one of the missions of his or her campaign and term in office. Other charged political situations may include a disagreement between federal and state regulatory agencies concerning who has authority to regulate site activities. There may also be disagreement between agencies regarding the interpretation of an established set of regulations. The state, for example, may feel that the federal agency has not properly interpreted or imposed its own regulations. When this happens, the facility owners/operators may find themselves in the center of a political football game between these two agencies and not know with which regulations to comply.

2.3.3 Conflicting Regulations

It is an understatement to say that environmental regulations can be confusing. Many times, it is difficult to even identify which set of regulations apply to a site (i.e., those under CERCLA, RCRA, TSCA, or a combination thereof). In addition, within a particular set of regulations, there may be requirements which defy logic. For example, during the 1980s, the RCRA land ban regulations dictated that no RCRA hazardous waste could be placed in land disposal units. At the same time, these regulations indicated that once a material was classified a hazardous waste, it was always a hazardous waste, whether or not it was treated to remove toxic constituents. These regulations further dictated that environmental media, such as soil, contaminated by listed hazardous compounds were not defined as a hazardous waste until they were excavated for treatment. This created a situation for many facility owners where, from their point of view, the only logical thing to do was to leave the soil in place and untreated. The RCRA regulations have undergone modification, but prior to these modifications, it was clear that developing a cleanup strategy could be quite difficult. It should be noted that other such conflicts still exist and no doubt will exist in the future.

2.3.4   Uncertainty/Conflict with Traditional Engineering Processes

To illustrate the point of uncertainty/conflict with traditional engineering process, we will use as an example the CERCLA program. The remedial process as envisioned by the Superfund program is as follows.[22] The project begins with the development of objectives, budget, operating assumptions, and identification of applicable or relevant and appropriate regulations; the facility owner initiates a remedial investigation (RI) and prepares a feasibility study (FS) that compares remedial alternatives. The facility owner proposes and the EPA selects an alternative. The selected alternative is specified in the record of decision which is written by the EPA. A consulting firm will then design the selected alternative (remedial design). Construction firms then offer bids for implementing the remedial design (the remedial action) and then the design is implemented.

This process assumes that uncertainties can be reduced early in the life of a project. It is believed that the investigation and study phases can reduce uncertainty and thus reduce large expenses later. This process of study, design, and build evolved from traditional engineering services such as the designing of roads, bridges, chemical processing plants, etc. This process assumes that uncertainty is reduced to manageable levels during the RI and FS stages.[22]

When this traditional engineering approach is applied to hazardous waste remediation projects, it is assumed that technology can discriminate among remedial decisions (to the satisfaction of all the stakeholders) and provide defensible arguments that the selected remedy will reduce risk to acceptable levels in a cost-effective manner.[22]

The problem with the decision-making process when applied to the remediation of hazardous waste sites is that it may be based on a number of false assumptions about the efficacy of the technology. These assumptions include the following:[22]

- At the RI/FS stage, it is possible to determine and define the risks posed by a site.
- Risks can be reduced to acceptable levels through the application of remedial technology.
- It is possible to determine what acceptable risk means to the various stakeholders.
- It can be proved that, at any given hazardous waste site, acceptable cleanup/ risk levels have been achieved.

These assumptions may be invalidated by the high degree of uncertainty associated with:[22]

- The health-based risk assessment process
- The subsurface environment
- Site investigation protocol
- The remediation technologies
- Site closure approaches

2.3.4.1 Health-Based Risk Assessment Process

The purpose of the health-based risk assessment process is to establish cleanup targets based on the toxicity of the compounds of concern and concentrations to which individuals may be exposed. In determining dose–response relationships for suspected carcinogens, it is assumed that:[22]

- Adverse effects of a suspected carcinogen on animals are predictive of effects of that chemical on humans.
- Adverse human effects from a suspected carcinogen at high doses are predictive of adverse human effects at low doses.
- A human's dose response to a suspected carcinogen is linear; thus the effects at low doses can be extrapolated from the effects at high doses.
- There is no concentration of a suspected carcinogen below which no detrimental effects occur.

A report by the National Research Council in 1983 indicated that these assumptions have not been and, at present, cannot be proved.

2.3.4.2 The Subsurface Environment

The subsurface environment is usually not well understood by the various project stakeholders. Few individuals have recognized the extent to which uncertainty exists and the effect it has on the remedial process. The subsurface medium is highly variable (heterogeneous), and small changes (either natural or man-made) in the physical or chemical nature of the subsurface can have major impacts on subsurface flow of water and contaminants and may lead to large variances between model predictions and observed results.

The range of these heterogeneities at a given site can be very broad. For example, permeability (a primary property used to quantify the ability of a fluid to pass through a medium) of the subsurface materials at a site can span a range of 10 billion units for common subsurface materials like clay, sand, and gravel (see Figure 2.4). It is possible to find a combination of these materials at any given site, each having an effect on the flow of groundwater and transport of contaminants.

Human thinking and experience are not well suited for comprehending such a wide range of variability. For comparison, the range in variability of the common physical property density is much smaller than the range in variability of the permeability of common geologic materials. For example, the density of gold (one of the heaviest known elements) is only about 50 times the density of water or less than two orders of magnitude more dense. This is a small range compared to the over ten-order-of-magnitude difference in permeability among common geologic materials.

Despite major advances in science, consultants are presently unable to definitively characterize the subsurface with current investigative techniques. Because the subsurface environment is so heterogeneous, it is that much more important to investigate a site in sufficient detail to remove uncertainties about the nature of the subsurface environment. Sampling to remove such uncertainties is, for the most part, beyond economic feasibility, regardless of the site. The average remedial investigation generally includes sampling a statistically very small portion of the subsurface, in many cases sampling less than one-millionth of the site by volume.[22]

Locating all significant contaminants is essential to developing engineering solutions to these problems. Yet, a quantity as small as 5 gallons can contaminate an aquifer that contains a billion gallons of water to above the 5-ppb criterion established for compounds such as benzene. Therefore, even the most extensive sampling efforts can miss significant quantities of contaminants.

2.3.4.3 Technology for Remediation

In order to develop appropriate engineering solutions for the problems at hazardous waste sites, reliable performance information is required on all remedial technologies. The fact is, most technologies utilized for the treatment of affected soils and groundwater have been adapted from chemical-processing technologies which were originally developed to process streams of known volume and constant composition.[22] When dealing with soil treatment, in particular, the chemical and physical composition of the medium can be extremely variable. In addition, since the environmental remediation industry is relatively young, reliable performance data are rather limited. Finally, because of the high degree of variability from site to site, the success of a technology at a particular site is not necessarily an indicator of success at another site.

2.3.5 Unrealistic Specifications

Cleanup targets for hazardous waste site remediation projects can be seen as the equivalent of engineering specifications that may be associated with other engineering projects such as manufacturing soap. Whereas Ivory Soap's claim to fame is that it is 99.44% pure, this specification comes nowhere near the drinking water standard. For example, achieving the MCL for benzene contamination in groundwater would make groundwater 99.9999995% pure. This represents a 5-ppb quality goal, which is as difficult to comprehend as the range in permeability values for common geologic materials. An analogy to visualize 5 ppb would be to compare the starting lineup of the Chinese National basketball team to the whole population of mainland China (one billion people).

2.3.6 Effect of Complicating Factors on Remedy Selection

When the Superfund remediation process was developed (as well as the processes for many other environmental programs), it was assumed that the process of assessing risk, determining acceptable risk levels, and successfully implementing remediation would be relatively straightforward.[22] Figure 2.5 presents a graphic representation of this process. For many sites, industry experience has indicated that because of the high degree of uncertainty, additional study does not necessarily reduce uncertainty, as in traditional engineering.

In remediation engineering, uncertainty is high and remains so long into the process. Figure 2.6 qualitatively compares the subsurface hazardous waste engineering process with traditional engineering. The complicating factors combine to make it difficult to use current scientific and engineering techniques to locate all significant contaminants, accurately determine risks associated with them, determine what constitutes acceptable risk, and develop a remedy to reach acceptable risk levels.[22] With such a high degree of uncertainty, there is considerable opportunity for each stakeholder to develop different assumptions and interpretations based on differing objectives.[22]

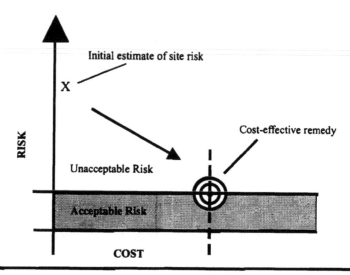

FIGURE 2.5   Risk assessment and remedy selection as assumed by the Superfund program. (From Wallace, W.A., New paradigms for hazardous waste engineering, *Remediation,* 1, 419, 1991 [original publisher Executive Enterprises Publications Co., New York)]. Reprinted by permission of John Wiley & Sons, Inc.)

In addition, it can be rather easy for any stakeholder to develop a set of credible assumptions to dispute the claims of another.[22] The final accepted assumptions and alternatives can have a significant impact on the overall cost of a project, sometimes on the order of tens of millions of dollars. Figure 2.7 qualitatively depicts the inability of the process to discriminate among alternatives.

Table 2.3 quantitatively demonstrates the above statements. This table compares the costs of treating soil with PCB contamination at an operating facility. For this particular site, a total of 13 alternatives were considered during the feasibility study. The alternatives included everything from no further action at the site (except for continued monitoring) to excavation and off-site disposal of all soil with PCB concentrations in excess of 25 ppm. Three on-site treatment technologies were retained by the study based on an extensive screening process as well as bench-scale experiments:

- Potassium polyethylene glycolate (KPEG) dechlorination
- On-site incineration
- Stabilization

The cost to implement the treatment technologies at different cleanup levels was evaluated. Finally, three facility operating scenarios were considered. Sce-

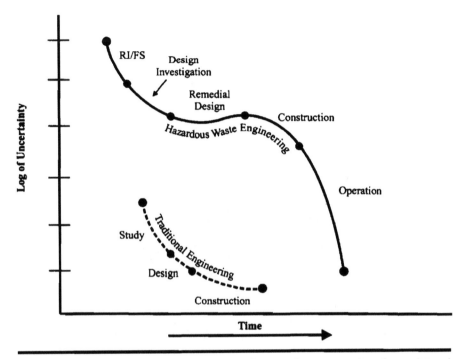

FIGURE 2.6 Qualitative cost comparison of uncertainty versus time for traditional and hazardous waste engineering. (From Wallace, W.A., New paradigms for hazardous waste engineering, *Remediation*, 1, 419, 1991 [original publisher Executive Enterprises Publications Co., New York]. Reprinted by permission of John Wiley & Sons, Inc.)

nario 1 was that the facility would continue its operations during cleanup. Scenario 2 was that the site would discontinue operations during cleanup and continue operations after cleanup. Scenario 3 was to remove the facility from service during and after cleanup. Since some of the alternatives required long-term monitoring and maintenance, cost comparisons were done on a 30-year net present worth basis. The net present worth cost for the various alternatives ranged from approximately $550,000 to approximately $44.6 million. The alternative proposed by the potentially responsible party was 4C under scenario 3 (Table 2.3), which had a net present cost of approximately $8.4 million.

After the required regulatory and public scrutiny, the proposed alternative was selected (however not without dissenting opinions). It should be noted that a strong case could be made for nearly every one of the alternatives, including the no-action alternative.

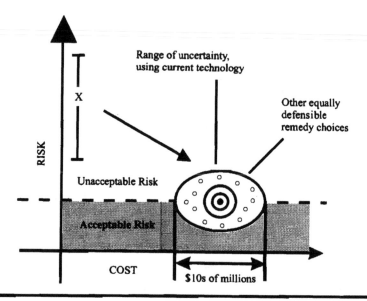

FIGURE 2.7   The technical reality of risk assessment and remedy selection. (From Wallace, W.A., New paradigms for hazardous waste engineering, *Remediation*, 1, 419, 1991 [original publisher Executive Enterprises Publications Co., New York]. Reprinted by permission of John Wiley & Sons, Inc.)

2.4   SCOPE MANAGEMENT

The previous sections on complicating factors presented a rather bleak picture in regard to the cleanup of hazardous waste sites. Yet there are ways of dealing with the high levels of uncertainty, managing stakeholders, and keeping costs to a minimum. The remainder of this chapter opens the door to techniques that can be used to keep overall project costs down while at the same time providing the greatest overall benefit to human health and the environment. The remaining chapters of this book provide the details of implementing these techniques. Note that these suggestions are written from the viewpoint of site owners and operators, yet they apply equally to the service companies trying to serve the needs of these owner/operators

2.4.1   Become and Remain Proactive

Because addressing environmental issues can have a negative impact on overall company profitability, many companies in the past have preferred to take a

TABLE 2.3   Quantitative Cost Comparison of PCB Cleanup Options at an Active CERCLA Site

	Present worth cost (%)		
Alternative	Scenario 1	Scenario 2	Scenario 3
1. No action	550,000	550,000	550,000
2. Institutional control	560,000	560,000	560,000
3. Soil cover containment	*	*	11,300,000
4. Treatment of soil with PCB concentration greater than 500 ppm			
A. KPEG	24,230,000	18,670,000	11,100,000
B. On-site incineration	*	*	14,330,000
C. Stabilization	*	*	8,410,000
5. Treatment of soil with PCB concentration greater than 500 ppm and containment of soil with PCB concentration between 25 and 500 ppm			
A. KPEG	*	*	15,400,000
B. On-site incineration	*	*	18,620,000
C. Stabilization	*	*	12,540,000
6. Treatment of soil with PCB concentration greater than 25 ppm			
A. KPEG	39,800,000	31,350,000	24,420,000
B. On-site incineration	44,600,000	36,090,000	29,170,000
C. Stabilization	31,750,000	23,190,000	16,270,000
7. Off-site TSCA landfill disposal of all soil with PCB concentration greater than 25 ppm	44,360,000	33,570,000	26,800,000

* Cannot be implemented under this scenario due to site constraints.

passive role in addressing environmental matters. Such companies chose to not properly manage released hazardous chemicals until forced by a regulatory agency to investigate the release. Once this happens, the owner/operator is forced into a reactive mode with little time to evaluate and hire legal and technical counsel. This leaves little time for technical and legal counsel to assist in the development of project and environmental management strategies, particularly to the level of detail prescribed in this book

However, even if the project was initiated unexpectedly, the proactive philosophy should be adopted as soon as possible. With the stakes in terms of cost and liability so high, the facility owner cannot afford to have the project driven by regulatory agencies, the media, and the public, who by and large are not as concerned about the overall final cost or whether more work than was necessary was performed.

2.4.2 Develop a Strategic Plan

In nearly every endeavor undertaken by industry, a strategic plan of one type or another is developed to address uncertainty, manage project risk, and develop responses that will lead to project success. Yet, experience has shown that when environmental projects are encountered, such prudent planning is replaced by a "let's go do it" attitude without considering the long-term effects of initial actions. Activities are then undertaken that commonly generate more questions than they answer.

In order to develop a strategic plan, facility owners must be able to identify appropriate project objectives, risks in meeting these objectives, the costs involved, and responses for reducing these risks. The issues that must be considered in developing the strategic plan are multidisciplinary and will require the efforts of a multidisciplinary team to ensure that all of them are identified and addressed.

2.4.3 Develop Long-Term Objectives

When faced with an environmental remediation project, the complexities can be so overwhelming that it may be difficult to obtain agreement on objectives for the strategic plan. Yet, what is desired by all is regulatory site closure at the lowest possible cost and minimum future liability. Regulatory site closure is defined as regulatory approval and agreement that the site no longer presents a threat to human health and the environment.

In order to obtain regulatory site closure, the project team must first obtain regulatory approval of a remedial action plan and, second, investigate and remediate the site to the point where technically defensible documentation may be developed which supports closure.

2.4.4 Focus on Contaminant Mass

Experience indicates that projects that focus on removing the greatest amount of mass of contaminants from the environment will achieve the greatest risk reduction with the lowest overall cost. This concept might seem obvious, but it is typically forgotten as a result of the way in which regulations were historically written and implemented.

Many of the cleanup standards that have been established by the regulatory agencies are for the protection of groundwater. Groundwater represents one of the major mechanisms for the transport of contaminants away from a site and for exposure of individuals. Cleanup standards are commonly written in terms of concentration (i.e., parts per billion). Therefore, when a facility owner discovers that the groundwater under the plant exceeds one of these cleanup standards, a project to address groundwater cleanup may be undertaken without considering mass distribution of contaminants.

Figure 2.8 indicates that when loss of an organic chemical occurs in the environment, as much as 60 to 100% of the lost chemical can become adsorbed to the subsurface soils, Another 10 to 40% can be found as liquid-phase hydro-carbons (either floating on the water table or sinking to the base of the aquifer). Only 10 to 20% might be found dissolved in the water. The soil may act as a continual source for contaminants to impact groundwater. A program designed to address only one phase of impact, particularly if it is the phase that contains the least amount of mass, can be extremely costly and of long duration since the other phases may act as a continual source of impact to the groundwater. By cleaning up the phase that contains the greatest mass of contaminant, for ex-ample the soil, the groundwater program will be greatly shortened and the en-vironment will be better protected than if the soil had not been addressed. This fact must be considered when developing the strategic plan.

2.4.5   Focus on Risk

The uncertainties associated with the subsurface environment in many cases may make it impossible to achieve general (nonsite-specific) cleanup standards since the standards, such as MCLs, are so low. However, well-researched and detailed risk assessments can indicate that general cleanup standards are overly conservative. This risk assessment can be utilized to justify leaving residual contamination in situations where the contaminant cannot be removed by the best available technology approach or the best available technology is economi-cally unfeasible.

2.4.6   Develop Short-Term Objectives

Once the long-term objectives are established and the concept of focusing on mass of contaminant and on risk is accepted, the project team is in a position to begin developing short-term objectives. These objectives typically involve char-acterizing the site both physically and analytically so that enough data are gath-ered to evaluate remedial alternatives and develop initial remedial designs. In addition, a short-term objective should include the implementation of some form

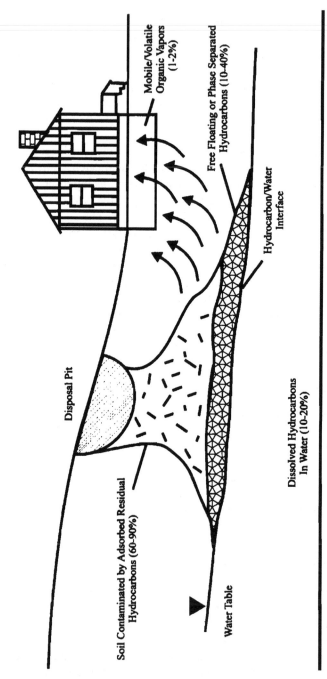

FIGURE 2.8 Generalized distribution of organic chemicals subsequent to a subsurface release. (From Yaniga, P.M. and Smith, W., Aquifer restoration: in situ treatment and removal of organic and inorganic compounds, in *Proceedings Groundwater Contamination and Reclamation*, Schmidt, K.D., Ed., American Water Resources Association, Tucson, AZ, 1985, 150. Reprinted by permission of the American Water Resources Association.)

of remedial action for those areas of the site where it is certain that some form of remedial action will be required. This type of remedial action could include activities such as the removal and proper disposal of waste drums, the excavation of leaking underground storage tanks, or the use of pumping wells to remove liquid-phase contamination from the surface of the water table.

Industry experience in dealing with the cleanup of hazardous wastes sites at numerous facilities and under a multitude of regulations has led to the development of a generalized work breakdown, which is presented as Figure 2.9.

This work breakdown structure indicates first and foremost that these projects must be conducted in phases. These phases provide one of the mechanisms for dealing with uncertainty in that the results of previous phases can be analyzed and the activities in the following phases can be adjusted based on knowledge gained in the previous phases. Figure 2.9 indicates that interim remedial actions can be performed at the same time as actions to physically and analytically characterize the site.

The physical characterization pertains to the development of data regarding the geology, hydrogeology, surface runoff, and other physical features. These physical qualities can affect the dispersion of contaminants into the environment. The analytical characterization provides data regarding the types of compounds that have been released into the environment, the medium which has been affected, and the concentration and mass of the contaminants in each medium. These combined data provide information regarding the nature and extent of contamination.

With physical and chemical characteristics of the contaminated medium defined, a risk assessment to establish cleanup objectives can be performed. Upon completion of the risk assessment, an evaluation of remedial alternatives (also known as a corrective measures study or a feasibility study) may be performed. The combination of the risk assessment/feasibility study may indicate that more information is required, that no action is required and that a request for closure should be developed and defended, or that enough data have been gathered to begin design and implementation of remedial actions.

2.4.7 Utilize Modern Economic Risk Management Processes

Modern risk management processes such as quantitative decision analysis and stochastic project modeling should be included in all parts of the strategic planning. These technique are used in many industries to evaluate alternative approaches when there is great uncertainty. The petroleum exploration and production industry commonly uses the economic decision tree approach for planning their projects. Chapter 10 addresses modern risk management processes in great detail.

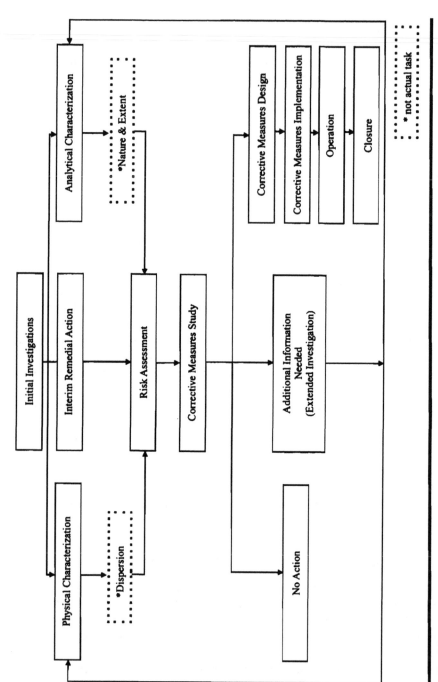

FIGURE 2.9   Generalized work breakdown structure.

2.4.8 Implement Project Management Controls

Once the strategic plan has been identified, short- and long-term goals and objectives and selected pathways of approach, budgets, and schedules can be developed and tracked (one phase at a time) utilizing standard project management techniques. These techniques, including their application in the environmental remediation industry, are the focus of Chapters 4 through 8. Standard project management practices must then be utilized to control costs at each and every phase (see Chapter 9).

2.4.9 Steps for Project Completion

In light of the previous recommendations, the steps for cost-effective project completion can be summarized as follows:

- Utilize strategic planning to plan overall project scope, direction, and purpose.
- Conduct the project in phases.
- Perform interim remedial actions while performing physical and analytical site characterization.
- Focus remediation on the media that contain the largest mass of contaminants.
- Perform a risk assessment to establish cleanup goals and to justify leaving small residual amounts of contamination.
- Perform a feasibility study designed to select an appropriate combination of treatment and removal options.
- Implement remedial technologies.
- Document total mass of contaminants removed from all media.
- Develop a technically defensible closure plan.

After discussing some of the basics concepts of modern project management in Chapter 3, we will be ready to see how these techniques can be utilized to carry out the above recommendations.

REFERENCES

1. Lion Technology, *Environmental Regulations Course,* Vol. 1, ERO4097, Lion Technology, Lafayette, NJ, 1997, 1-1.
2. Lion Technology, *Environmental Regulations Course*, Vol. 1, ERO4097, Lion Technology, Lafayette, NJ, 1997, 1-2.
3. Lion Technology, *Environmental Regulations Course,* Vol. 1, ERO4097, Lion Technology, Lafayette, NJ, 1997, 1-11.

3a. Jain, R.A., Environmental legislation & regulations, in *Standard Handbook of Environmental Engineering*, Corbitt, R.A., Ed., McGraw-Hill, New York, 1990, 2.1.

 4. Jain, R.A., Environmental legislation & regulations, in *Standard Handbook of Environmental Engineering*, Corbitt, R.A., Ed., McGraw-Hill, New York, 1990, 2.2.

 5. Schultze, C.L., *The Public Use of Private Interest*, The Brookings Institution, Washington, D.C., 1977, 32.

 6. Jain, R.A., Environmental legislation & regulations, in *Standard Handbook of Environmental Engineering*, Corbitt, R.A., Ed., McGraw-Hill, New York, 1990, 2.3.

 7. Rogoff, M.J., Status of state brownfield programs: a comparison of enabling legislation, *Remediation Management*, Second Quarter, 14, 1997.

 8. Jain, R.A., Environmental legislation & regulations, in *Standard Handbook of Environmental Engineering*, Corbitt, R.A., Ed., McGraw Hill, New York, 1990, 2.5.

 9. Lion Technology, *Environmental Regulations Course*, Vol. 1, ERO4097, Lion Technology, Lafayette, NJ, 1997, 7-1.

10. Jain, R.A., Environmental legislation and regulations, in *Standard Handbook of Environmental Engineering*, Corbitt, R.A., Ed., McGraw-Hill, New York, 1990, 2.10.

11. Jain, R.A., Environmental legislation and regulations, in *Standard Handbook of Environmental Engineering*, Corbitt, R.A., Ed., McGraw-Hill, New York, 1990, 2.11.

12. Abron, L.A. and Corbitt, R.A., Air and water quality standards, in *Standard Handbook of Environmental Engineering*, Corbitt R.A., Ed., McGraw-Hill, New York, 1990, 3.7.

13. Jain, R.A., Environmental legislation and regulations, in *Standard Handbook of Environmental Engineering*, Corbitt, R.A., Ed., McGraw-Hill, New York, 1990, 2.14.

14. Sutherson, S.S., *Remediation Engineering: Design Concepts*, Lewis Publishers, Boca Raton, FL, 1997, 4.

15. McCain, W.D., Jr., *The Properties of Petroleum Fluids*, Pennwell Publishing, Tulsa, OK, 1973, 45.

16. Pankow, J.F. and Cherry, J.A., *Dense Chlorinated Solvents and Other DNAPLS in Groundwater*, Waterloo Press, Portland, OR, 1996.

17. Freeze, R.A. and Cherry, J.A., *Groundwater*, Prentice-Hall, Englewood Cliffs, NJ, 1979, 44.

18. Lindeburg, M.R., *Engineering-in-Training Reference Manual*, 8th ed., Professional Publications, Belmont, CA, 1992, 14-11.

19. Sutherson, S.S., *Remediation Engineering: Design Concepts*, Lewis Publishers, Boca Raton, FL, 1997, 12.

20. Driscoll, F.G., *Groundwater and Wells*, 2nd ed., Johnson Division, St. Paul, MN, 1986, 74.

21. Freeze, R.A. and Cherry, J.A., *Groundwater*, Prentice-Hall, Englewood Cliffs, NJ, 1979, 27.

22. Wallace, W.A., New paradigms for hazardous waste engineering, *Remediation*, 1, 419, 1991.

General Project Management Concepts

3

As stated in Chapter 1, the focus of this book is the pragmatic application of modern project management techniques to the environmental remediation industry. As such, this book is more concerned with practice than theory. However, it is neither possible nor desirable to avoid a discussion of theory altogether. In fact, an understanding of the fundamental concepts of project management theory can assist the project manager in gaining an in-depth understanding of the project at hand — and make sound decisions based on that understanding. Therefore, the purpose of this chapter is to provide a general understanding and appreciation of the art and science of project management, as well as the fundamental concepts of project management theory. Although discussions of project management theory can be found throughout this book, the most extensive discussion is presented in this chapter. As the theoretical concepts are presented, every attempt has been made to provide practical examples of how theory can be applied to the actual management of environmental remediation projects.

In order to summarize the significant concepts of project management theory, it is necessary and appropriate to draw upon the works of its most ardent proponents. These include educators such as Dr. Harold Kerzner of Baldwin-Wallace College and Dr. David I. Cleland of the University of Pittsburgh, as well as the variety of materials developed by the Project Management Institute (PMI). It is strongly recommended that readers interested in a deeper understanding of project management theory access the works of these as well as other authors and become members of the local chapter of the PMI.

3.1 WHY PROJECT MANAGEMENT?

The essential feature of any project is to bring about change. Projects are initiated to create products or situations that are more desirable than those currently in place. Simply stated, project management is about making things happen, accomplishing change. It is important to note the use of the word "management." It indicates that project managers accomplish change through management methods such as planning, directing, coordinating, and integrating the activities of others (the project team). Although this is a management rather than a technical role, project managers are most effective when they have a functional, if not detailed, knowledge of the various facets of the industries in which they are operating (i.e., project managers need to understand or have a working knowledge of the technical field).

The resources available to accomplish change are usually, if not always, limited. Since resources are limited, project managers are always attempting to accomplish the maximum amount of change using the minimum amount of resources. Project management, then, can be further understood as attempting to do more with less.

This "doing more with less" approach to problem solving has been referred to by the great scientist, engineer, architect, and inventor R. Buckminster Fuller as the process of ephemeralization. This process is firmly in place, for instance, in the computer, communications, and transportation industries, where technologies are becoming faster, more powerful, and more efficient — while using fewer resources. Finding new ways to accomplish more useful work while utilizing fewer resources is necessary for the competitive survival of all companies.

Our present technological and competitive environment is characterized by rapid, accelerating change. Modern project management processes represent a sound approach for coping with a changing environment by allowing more work to be accomplished with fewer resources. The additional work is accomplished by *working smarter, not harder.*

3.2 BRIEF HISTORY OF PROJECT MANAGEMENT

The origins of project management are rooted in antiquity.[1] The Egyptian pyramids, the Great Wall of China, Roman buildings and roads, and shipbuilding are all examples of the use of project management.[1] Although no one can claim credit for its invention, the actual formal beginnings of modern project management are often traced to the U.S. ballistic missile program or space program.[1] At the present time, project management has reached a high degree of maturity and is widely used in industrial, educational, governmental, military, and now environmental circles. A distinct field of literature has emerged which deals with project management in contemporary organizations.

3.3   PROJECT MANAGEMENT IS A PROFESSION

The development of project management as a profession in its own right is reflected in the literature, the activities of professional organizations, and the project management degree programs offered by colleges and universities.[1] Since the early 1960s, over 100 books have been published on the various aspects of project management.[1] Additionally, hundreds of articles on project management have appeared in management and technical publications.

Professional associations such as the PMI and the International Project Management Association hold seminars and symposia each year at which members present papers regarding their special areas of interest in project management.

The PMI was founded in 1969 and as of the end of 1997 had in excess of 30,000 members worldwide. In addition, it has certified over 2000 individuals as Project Management Professionals. According to PMI records, at the end of 1997, 36 colleges and universities offered graduate-level degrees in project management. Given the growth trends of the 1990s, this list is expected increase.

Although project management has come into its own as a profession, it should be understood that project management is an evolving field of study, whose different concepts and practices are at times widely accepted or, conversely, subject to great debate.

3.4   ENVIRONMENTAL PROJECT MANAGEMENT

In order to understand the theory and practice of project management, one must begin with a definition of a project. Once a project has been defined, it is possible to move further — to define *project management* and, ultimately, to use these definitions to describe environmental project management.

3.4.1   Just What Is a Project?

The definition of a project has been the subject of considerable debate among the practitioners of the profession. Evidence as to the evolvement of the definition is provided by the fact that the PMI published its current definition of a project in the 1996 edition of the *Project Management Body of Knowledge* (PMBOK). This definition updated the definition provided in the June 1993 issue of the *Project Management Journal,* which is also published by the PMI. This definition in turn updated a previous definition which appeared in the PMI's March 1987 edition of the PMBOK. It should be noted that the PMBOK serves as the official knowledge standard and defines the minimum knowledge that an individual must possess to be considered a recognized Project Management Professional.

Perhaps one of the things that has made it difficult to define a project is that many individuals feel that the definition of a project can be accepted as self-evident ("I know a project when I see one"). Yet different individuals will often disagree, when provided with examples, over whether or not certain activities represent a project.

In the next several pages, we will review the various definitions of a project that have existed in the recognized literature over time as well as the current definition in the PMI PMBOK. These definitions are utilized to derive a definition that is suitable for the environmental industry and reflective of the recommended process for planning environmental remediation projects. This process is presented at the end of this chapter and expounded in Chapters 4 through 9.

This is not to say that the earlier definitions of a project are in error; these definitions severed the needs of industry and the stage of maturity of the project management profession at the time they were provided. In addition, the current definition provided by PMI's PMBOK is considered correct. The PMI serves the needs of a wide body of industries, and its definition must be accurate and general enough to represent of all its members. Here, we have the luxury of focusing on the needs of a particular industry.

The various definitions of a project that have appeared over time include the following:

1. "Any undertaking that has definite, final objectives representing specific values to be used in the satisfaction of some need or desire."[2]
2. "A project can be considered to be any series of activities and tasks that:[3]
   - Have a specific objective to be completed within certain specifications;
   - Have defined start and end dates;
   - Have funding limits;
   - Consume resources (i.e., money, people, equipment)."[3]
3. "...Simply a cluster of activities that is relatively separate and clear cut. A project has a distinct mission and a clear termination point."[4]
4. "A project is any undertaking with a defined starting point and defined objectives by which completion is identified. In practice, projects depend on finite limited resources by which objectives are to be accomplished."[5]
5. "Resource consuming sets of activities through which strategies are implemented and goals are pursued."[6]

6. "A temporary process undertaken to create one or a few units of a unique product or service whose attributes are progressively elaborated."[7]
7. "A temporary endeavor undertaken to create a new project or service."[8]

The first five definitions use the term objectives, or similar terms such as goals or distinct mission, to describe one of the attributes of a project. The seventh definition, which is the most recent one provided by the PMI, does not speak in direct terms of goals or objectives. However, the need for a distinct objective is alluded to by reference to the creation of a new product or service. Since all of the above definitions suggest the need for well- defined objectives in order to establish a project, this term will be included in the definition derived for the environmental remediation industry.

Definitions 2, 3, and 4 refer to a series, cluster, or set of activities as representing one of the attributes of a project. There are two reasons why this attribute of a project also appears appropriate for inclusion in the definition for the environmental remediation industry. The first reason is that environmental projects are typically performed in phases; inclusion of this attribute acknowledges these phases. The second reason is that dividing a project into a number of distinct manageable activities, as will be discussed later, is one of the desirable first steps in developing a project plan.

The second definition, provided by Kerzner, appears most comprehensive. This definition acknowledges that the activities must be completed in accordance with established specifications (scope), must have defined starting and ending dates (schedule), require funding limits (budget), and will consume resources (material, equipment, people). This definition is comprehensive in that it acknowledges the basic questions that are likely to be asked by any customer faced with an environmental project:

- What will I have to do?
- How long will it take?
- How much is it going to cost?

Kerzner's definition also acknowledges another question which must be answered by a company proposing to perform a project to satisfy a customer's needs: What resources will be required (both internal and external) in order to complete the project? The fact that project activities consume resources is also indicated by definitions 4 and 5. (Indeed, definition 4 further elaborates on this point, noting that the resources are in limited supply. In other words, there are only so many technicians, geologists, engineers, etc. available to work on a particular project.)

There is one remaining attribute of a project that is found in definitions 6 and 7 when they refer to a project as "creating a unique product." This is an important reference because it calls attention to the difference between operations management and project management. Repetitive operations require traditional forms of management rather than project management. For example, once a chemical manufacturing plant has been constructed, management's function is to keep the plant operating efficiently, while it continues to perform the same repetitive task. A more traditional hierarchical type of management is appropriate for operating a chemical plant. Note that construction of the plant would require project management, but operating it would not. One of the reasons that the environmental remediation industry is project driven is that each site is unique in surface operations and the subsurface conditions. In addition, one does not expect to clean up the same site twice. The uniqueness of each environmental remediation job, therefore, calls for a project management approach to its planning, implementation, and control. In consideration of the above discussion, the definition of a project considered most appropriate to the needs of the environmental remediation industry is as follows.

A project is a unique undertaking that consists of:

- A specific objective
- A series of tasks
- Defined scope and specifications
- Schedule for completion
- Budget
- Resource consumption

3.4.2 Project Management Definition

Now that a project has been defined, we can address the term "project management." Project management can mean different things to different people. Quite often, people misunderstand the concept because they have ongoing projects and feel that they are using project management to control these activities.[9] According to Kerzner, in such a case, the following may be considered a definition of project management:[9]

> Project management is the art of creating the illusion that any outcome
> is the result of a series of predetermined, deliberate acts when in fact
> it was dumb luck.

Regardless of this facetious definition by Kerzner, and people's misunderstandings about project management in general, there appears to be less controversy in the literature regarding the definition of project management. Perhaps this is because the functions of classical management are fairly well defined.

Classical management is considered to have five functions or principles:[10]

- Planning
- Organizing
- Staffing
- Controlling
- Directing

The one difference between project management and classical management is that typically the project manager in project management does not staff the project; in most companies that use formalized project management, staffing is considered a function of operations management. Although the project manager has the right to request specific resources, including staff, the final decision on committing resources rests with the operations managers.

Given our working definition of a project and our understanding of the classical management functions, we can arrive at the following definition:

> Project management is the art and science of planning, organizing, integrating, directing, and controlling all committed resources — throughout the life of a project — to achieve the predetermined objectives of scope, quality, time, cost, and customer satisfaction.

3.4.3 Environmental Project Management

In addition to performing in accordance with customer satisfaction, as stated in the above definition of project management, environmental remediation projects must be performed in accordance with environmental regulatory requirements as well (e.g., RCRA, CERCLA, TSCA, UST). Furthermore, environmental project activities must be performed within the appropriate health and safety requirements (in particular, those outlined in 29 CFR 1910.120). Finally, environmental remediation undertakings are likely to generate a great deal of public interest and concern. Therefore, these projects must be performed while maintaining good public relations.

Figure 3.1 is a pictorial representation of environmental project management. This figure, adapted from Kerzner, shows that environmental project management involves the following elements:

- Managing or controlling committed company resources on a given project
- Within the constraints of time, cost, and performance/technology
- And within the further constraints of health and safety requirements, regulatory requirements, good customer relations, and good public relations

For those project managers who work for remediation service companies, successful project management entails completing a project within these constraints while **earning a profit** for their employers.

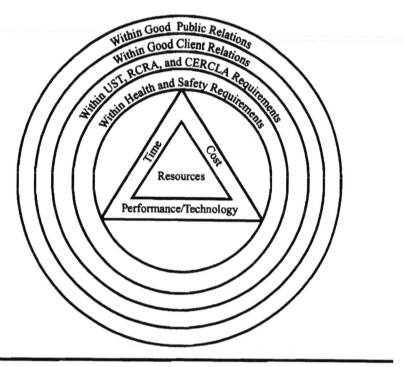

FIGURE 3.1   Environmental consulting project management. (From Kerzner, H., *Project Management: A Systems Approach to Planning, Scheduling, and Controlling,* Van Nostrand Reinhold, New York, 1984. Reprinted by permission of John Wiley & Sons, Inc.)

3.5   PROJECT CRITERIA

Having defined a project as well as environmental project management, we can now turn our attention to the criteria of a project. These criteria allow the project manager to take a detailed, pragmatic look at the planning, implementation, and control of a project.

The criteria for successful project management are as follows:

- **Written long-term and short-term objectives** that have been agreed upon by the customer and the remediation service company while considering all stakeholders and drivers
- **A written scope of work,** which both defines and limits the extent to which the tasks will be performed, including written specifications, assumptions, terms and conditions, etc.

- **A work breakdown structure** which clearly outlines the scope of work and separates the scope of work into a series of manageable tasks
- **A detailed schedule** which indicates the estimated duration of all tasks and their order of occurrence
- **An identified project team** with team members assigned to each of the identified tasks and one individual (task manager) held responsible for the management of that task
- **A project budget** determined by assigning resources (human, material, etc.) to each of the tasks identified by the work breakdown structure

3.6   PROJECT MANAGEMENT RESOURCES

Since projects consume resources in reaching their objectives, there is a need to control these resources within time, cost, and performance. Six primary resources are available to the project manager:[11]

- Money
- Personnel
- Equipment
- Materials
- Information/technology
- Facilities

Note that the project manager does not actually control any of these resources, with the possible exception of money (i.e., the project budget). Instead, resources are controlled by operations management (referred to as line management in some organizations). The project manager must, therefore, negotiate with the operations managers for all project resources.

There is another way of looking at this concept. When it is said that a project manager controls resources, what is really meant is that the project manager **controls** those resources which have been temporarily **loaned** by the operations manager. Since the operations managers have the ultimate authority over the resources, it should be obvious that successful project management depends on:[11]

- A good working relationship between the project manager and those operations managers who directly assign resources
- The ability of employees to report vertically to their operations managers at the same time they report horizontally to one or more project managers

3.7 PROJECT MANAGEMENT FUNCTIONS (KNOWLEDGE AREAS)

We can now look at the actual job functions of a project manager. Project management is a complex multidisciplinary profession which has considerable overlap into many disciplines and professions.[1] However, if project management is to be identified as a unique profession, there is a need to put boundaries on the body of knowledge. Attempts by the PMI to identify the unique project management body of knowledge originally resulted in eight project management functions. With the release of the 1996 PMBOK, the functions came to be known as "knowledge areas." In addition, a ninth knowledge area known as project integration was added. In the PMBOK, these knowledge areas describe project management and knowledge in terms of component processes.[12]

The nine project management knowledge areas are[12]

- Integration management
- Scope management
- Quality management
- Time management
- Cost management
- Risk management
- Human resources management
- Contract procurement management
- Communications management

This formal list of knowledge areas serves several purposes, including:

- Setting the boundaries for the project management profession
- Identifying the responsibilities of the project manager
- Defining the knowledge set and skills required by the project manager

An important point about these so-called knowledge areas is that although we can separate the various project management processes into such categories, for the purposes of communication and understanding, they are in fact interdigitated to a great degree. For example, when performing group planning regarding the scope of a project, the project manager is dealing with issues of communications management as well as human resources management.

3.7.1 Project Integration Management

Project integration management describes the processes to ensure that the various elements of a project are properly coordinated. These processes are project plan development, project plan execution, and overall change control. Since the focus of this book is on project planning for environmental remediation projects, it can be said that this is a book primarily about project integration management for the environmental remediation industry. In Chapters 4 through 8, a number of tools, techniques, and case study exercises for the development of project plans for environmental remediation projects are provided. Chapter 9, on the other hand, presents the concept of earned value, which can be used to monitor project execution and identify project variance (control change).

3.7.2 The Four Core Functions (Knowledge Areas)

The functions of scope, quality, time, and cost are known as the four core functions.[13] These core functions lead to specific objectives, which are integrated with one another within the project life cycle. (The project life cycle is discussed in detail in Section 3.8.) Together these functions form a frame of reference for a project, against which the success of the project can be measured.[13]

From the customer's point of view, the four core functions represent a set of requirements; from the project manager's perspective, they represent the parameters or constraints of a project. Either way, whether as objectives or parameters, these four functions — scope, quality, time, and cost — constitute the basics of project management.

3.7.2.1   Scope

The definition of the required output of a project is known as the project scope. Since the scope of a project must first be identified, developed, and later has a habit of changing during the rest of the project life cycle, this gives rise to *scope management*.[14]

3.7.2.2   Quality

For the products of a project to be considered satisfactory, certain standards of quality must be defined and then achieved. This leads to the need for *quality management*.[14]

3.7.2.3   Time

The life of a project is finite, which is to say that the time available for completion is limited. Note that in reality, time itself is inflexible. Therefore, the activities required for a project must be carefully planned and monitored if they are to be completed within the time available.[14] This is referred to as *time management.*

3.7.2.4   Cost

Within the duration of a project, the consumption of resources costs money. Therefore, the development of a budget through the assignment of resources and the monitoring of budget expenditures comes under the heading of *cost management.*[14]

3.7.3   Four Facilitating Functions

In addition to the four core functions, project management involves the functions of risk, human resources, contract/procurement, and information/communications. These are called the facilitating functions, because they are the means through which the objectives of the basic functions are achieved.[15] The facilitating functions are elaborated as follows.

3.7.3.1   Risk Management

As noted in Chapter 1, projects are launched for the purpose of implementing change. Even the process of project management itself is subject to considerable change throughout the course of the project life cycle (see Section 3.8). Because of the relative uniqueness of every project, and rapidly changing conditions, the final outcome of every project is always uncertain.[15] Uncertainty can be associated with probability and risk. Prudent management will take the steps to mitigate the possibility that requirements will not be met by reducing risk without compromising overall project objectives.[15]

In order to reduce project risk, a comprehensive understanding of the project must be developed at the outset, including detailed planning before the project is initiated. Steps to identify and mitigate potential risks are known as *risk management.*[16]

3.7.3.2   Human Resources Management

We know that projects are achieved through people and their respective skills and abilities. However, the number of people, and their types of skill, varies considerably during the course of a project; often, certain personnel are required on the project for only a very short time.[15] Normally, there will be a core project team, led by the project manager, but even this core team is required only temporarily.[15]

In this temporary setting, careful attention must be given to the assembly of people, their interactions, and their motivation to work together effectively, through a clear understanding of their respective roles and responsibilities.[15] This requires *human resources management.*

3.7.3.3   Contract/Procurement Management

The services of individuals working within an organization have traditionally been acquired through informal understandings. People negotiate daily with buyers, sellers, bosses, and co-workers in order to obtain formal and informal commitments. It is a common experience, therefore, that a major portion of a project manager's time must be spent in procuring people's commitment to the project objectives. Procurement, by the way, is the acquisition of something (anything) for money (or equivalent), including a job that pays a wage. Additionally, it includes carrying out the obligation to a successful conclusion.

Internal personnel are not enough for the successful completion of most projects. Services external to the organization (drilling, laboratory analysis, etc.) may be purchased through an "arm's length" contract, that is to say, a formal written arrangement vested in law.[15] Materials and equipment may also be contractually secured.

In project management, formalized techniques can be used to achieve commitments from internal personnel. These techniques involve written turnaround documents, which are issued to each project team member who will carry out a given task. A turnaround document describes the work to be performed, the level of quality, the deadline, and the number of billable hours and the equipment and materials that have been budgeted for completion of the task.

In effect, the employee receiving such a document has just been awarded a mini-project (task); the turnaround document represents the employee's working contract. In fairness to the employee who is issued such a document, ideally he or she has the opportunity to provide initial input on the amount of effort required for completion of the given task. Thus, the employee's input during the upfront project planning is desirable, but not always possible. Note that the turnaround documents are discussed in more detail in later sections of this book.

In total then, the commitment of goods and services to a customer, and to the project, as well as the administration of their conduct or delivery, are the responsibility of *contract/procurement management.*[17]

3.7.3.4   Information/Communications Management

Sound project management requires developing a plan, collecting information on the status of the work at any given time, comparing it to the plan, and, if necessary, taking corrective action.[17] However, this only works if people know and understand the plan and provide feedback.

Communication of the plan, collecting information regarding its execution, obtaining feedback from the personnel involved, and communicating the status of the project fall under the heading of *information/communications management.*

3.7.4 Project Management Functions and Boundaries

The project manager must interface with a variety of individuals during the planning and execution of a project. The better the project manager understands the skills offered by these individuals, the better he or she will be able to communicate with them. The most successful project managers are those who constantly work to gain knowledge and skills in a variety of disciplines. In other words, they work to become good generalists.

Since the project manager is carrying out a management role for what is often a technical project, the project manager's skills overlap into other disciplines. Because of this overlap, the project manager can be tempted to perform the functions of other disciplines. This is undesirable since it prevents those with expertise, responsibility, and authority from performing their respective roles. The three most common areas where the skills and knowledge of the project manager may overlap with those of other disciplines are[18]

- General management
- The technical areas in which the project is involved (in this case, the environmental remediation industry)
- Supporting disciplines and services

In the project management literature, this overlap is often depicted as a Venn diagram in which each circle represents a particular body of knowledge and the shaded area represents the overlaps (Figure 3.2).

In Figure 3.2, the overlap indicates that it is desirable for the project manager to have familiarity with, if not a strong working knowledge of, the other disciplines. However, the white space (the boundaries) indicates that the project manager should not perform the functions of these other disciplines.

If the project manager begins to assume the role of the other disciplines, it is often to the detriment of the project, since it takes time away from the defined project management functions (the nine knowledge areas [functions] defined by the PMI). In addition, when project managers go beyond their own tasks, it greatly limits the number of projects they can manage; this is because the project manager has, in fact, become one of the project's technical resources.

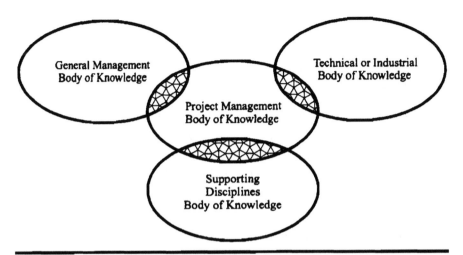

FIGURE 3.2 Project management boundaries and overlap. (From Wideman, R.M., *A Framework for Project and Program Management Integration*, Project Management Institute, Drexel Hill, PA, 1991. Reprinted from *A Guide to the Project Management Body of Knowledge* with permission of the Project Management Institute, 130 South State Road, Upper Darby, PA 19082, a worldwide organization of advancing the state-of-the-art in project management. Phone: (610) 734-3330, fax: (610) 734-3266.)

A general rule of thumb for the project manager is that project objectives are accomplished through the utilization of people who have unique technical abilities — from a variety of disciplines. The project manager's job is to coordinate and integrate the activities of these individuals (the project team), while maintaining a focus on the nine project management functions. Chapter 5 deals with assembling project teams and gaining the commitment of the individuals who comprise the teams.

The functions and types of knowledge typically associated with other professions are discussed here for the purpose of clarifying responsibilities. This is not meant to be an exhaustive discussion of all roles and responsibilities of other professionals, but only an attempt at clarification.

The types of functions and the body of knowledge typically associated with operations management (corporate, regional, territorial) include:[17]

- Business policy
- Business strategy
- Financial management
- Accounting

- Information systems
- Organizational behavior and development
- Staffing
- Personnel development
- Marketing strategy

Typical supporting disciplines, and their roles and responsibilities, include:

- Administration: Securing job numbers, establishing and maintaining project files, word processing
- Health and safety: Setting policies and procedures, developing health and safety plans
- Regulatory compliance: Providing regulatory guidance, interpretation, and strategy
- Contract administration: Reviewing contracts for acceptability of terminology, assisting in contract negotiations with customers
- Legal: Also involved in contract review, potentially involved in settling contract disputes and other legal issues
- Document production: Production assistance on proposals and lengthy reports
- Central engineering: Setting engineering policies, oversight review, developing standard operating procedures, technical support

The technical body of knowledge is perhaps the most diverse, particularly in the environmental remediation industry. The technical body of knowledge for the environmental remediation industry includes but is not limited to:

- Geology/hydrogeology
- Engineering, including all subdivisions (chemical, civil, electrical, mechanical, etc.)
- Toxicology (environmental risk assessments)
- Site investigation technology
- Remediation technology
- Regulatory requirements (RCRA, CERCLA, UST)
- Laboratory techniques and procedures
- Statistics
- Computer modeling and data manipulation
- Field technical procedures (investigation and remediation system installation)

The important point to recognize about these various bodies of knowledge and professions is that no one person can fully implement all these disciplines. Therefore, the project team must consist of a number of specialists — each supplying his or her respective body of knowledge. It is the project manager's responsibility to coordinate this team so that it functions efficiently and effectively to meet the needs of the project at hand.

3.8 PROJECT LIFE CYCLE

A project is by definition a unique undertaking, which means it has its own life cycle. The project management process, over the course of this life cycle, can be grouped into two sequential yet overlapping steps: planning and accomplishment. These sequential steps can be further broken down into four distinct phases in time through which any project passes: concept, development, implementation, and termination (finish). It should be noted that with the release of the 1996 PMBOK, the PMI defined project life cycle as a collection of phases whose number and names are determined by the needs of the performing organization. For the purposes of environmental remediation, the phases of concept, development, implementation, and termination are adequate.

In practice, on most projects, these four generic phases need to be broken down into greater detail, as shown in Figure 3.3. As indicated in the figure, each phase may be broken into a number of stages and further broken down into tasks. Note that while the phases are generic, the stages identified will be industry specific and the tasks identified are project specific.

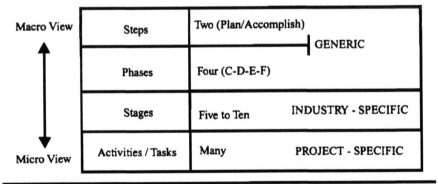

FIGURE 3.3 The anatomy of a project life cycle. (From Wideman, R.M., *A Framework for Project and Program Management Integration,* Project Management Institute, Drexel Hill, PA, 1991, iii-1. Reprinted from *A Guide to the Project Management Body of Knowledge* with permission of the Project Management Institute, 130 South State Road, Upper Darby, PA 19082, a worldwide organization of advancing the state-of-the-art in project management. Phone: (610) 734-3330, fax: (610) 734-3266.)

In an environmental remediation project, the industry-specific stages are often determined by the regulatory program which is driving the project. For example, if a project has been identified as falling under the domain of CERCLA, the industry-specific stages would correspond to Preliminary Assessment, Remedial Investigation, Risk Assessment, Feasibility Study, Remedy Selection, Record of Decision, Remedial Design, and Remedial Action.

As can be seen further, the industry-specific stages are further broken up into a number of manageable tasks, for the purposes of scheduling, budgeting, and controlling work. These tasks are obviously project specific. A detailed discussion of tasks specific to the environmental remediation industry is provided in Chapter 6.

3.8.1 Project Phase Activities

Although project phases are generic, that is, applicable to any sort of project, a discussion of the activities that typically occur in these phases is appropriate. Note that later in this chapter (Section 3.9.1, Table 3.2), life cycle stages are provided for environmental projects that are operated under various regulatory programs.

The **concept phase** is when a need is identified and background data are collected. At this point in a project, there is an attempt to determine if a problem exists. Is there a situation that can be changed for the better? Goals and objectives are developed during the concept phase, and initial plans are developed as well. It should be noted that although initial planning occurs in this phase, this does not mean that actual field work does not occur. The data gathering involved in the concept phase can be achieved by any means possible.

It should also be noted that the project can be terminated within this or any other phase of the project. For instance, an environmental project could be terminated at the concept phase because contamination was not found at the site.

In the **development phase**, more detailed planning occurs. By this point, the problem has been identified and a need has been established. The question at this point is: What is the best manner to go about solving this problem or satisfying our need? Economic studies are often performed in this phase, as well as an evaluation of alternatives. Detailed design will also be performed in this phase. At the **implementation phase**, construction activities often occur and the detailed design plans are implemented. The **termination phase** is the point where the results of the project are documented, resources are released, and responsibility for the newly developed product is transferred. For example, if the project were to build a new chemical manufacturing plant, it is at the termination phase that the plant would be placed in the hands of the new operators.

3.8.2 Significance of the Phases

Having provided an overview of the phases of a project life cycle, we can look at their relevance to various management concerns. The following section focuses on phases and the life cycle, in consideration of:

- Level of effort
- Cumulative costs
- Potential for reducing costs
- Escalating cost to fix or change
- Quality of information
- Risk versus amount at stake

3.8.2.1 Level of Effort

Throughout all phases of a project, the relationship between effort and time is quite distinctive.[20] (In this case, effort can be thought of as total labor hours per unit of time, or another way to think about effort is total cost incurred per unit of time.) To visualize this relationship, consider a curve of effort plotted against time.

Note in Figure 3.4 that the effort starts at zero, before the project has commenced, and ends at zero, after the project has been completed. Between these points, the effort over time curve has a typically characteristic profile, which may be likened to a pear sliced in half, resting flat-face down, with the stem at time zero.[20]

3.8.2.2 Cumulative Cost Over Time (the "S" Curve)

A curve can also be plotted as a progressive cumulative total (e.g., cumulative dollars over time). In this case, the curve looks like an "S," with the bottom tail of the "S" at time zero and the top tail corresponding to project completion.[21] (Note that project completion is determined by whatever effort and cost are necessary to reach that point.) Figure 3.5 presents a progressive, cumulative total curve, superimposed on the project management life cycle.

While Figure 3.5 presents an entire project life cycle, it should also be noted that the principle of the "S" curve can be equally applied to any of the tasks required as a component of a project. Experience within the environmental remediation industry in setting up numerous environmental projects using project management software has verified the existence of the characteristic shape of this "S" curve. The development of the cumulative cost over time curve is well within the capabilities of most automated project management information sys-

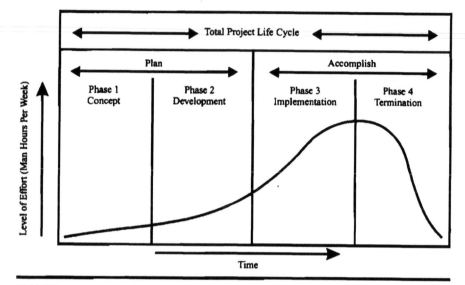

FIGURE 3.4   Project life cycle profile — level of effort. (From Wideman, R.M., *A Framework for Project and Program Management Integration*, Project Management Institute, Drexel Hill, PA, 1991, iii-7. Reprinted from *A Guide to the Project Management Body of Knowledge* with permission of the Project Management Institute, 130 South State Road, Upper Darby, PA 19082, a worldwide organization of advancing the state-of-the-art in project management. Phone: (610) 734-3330, fax: (610) 734-3266.)

tems. In fact, this curve forms the basis of a project schedule and tracking technique known as earned value analysis (discussed in detail in Chapter 9).

3.8.2.3   *Project Manager's Level of Effort Curve*

The project manager's level of effort curve (Figure 3.6) is significantly different from the level of effort curve for the project as a whole. In fact, the project manager's level of effort curve can be seen as the exact reverse or mirror image of the curve for the entire project.

The project manager's level of effort curve indicates that the level of effort is highest during the initial phases of a project. This is because the project manager must spend a significant amount of time in developing the project plan. Once the project moves into implementation, the project manager's burden will be reduced or shifted to task managers.

Figure 3.6 shows an increase in effort at the very end of the project. This increase signifies the fact that the project manager will become very involved at the end of the project in bringing about closure. This could include developing

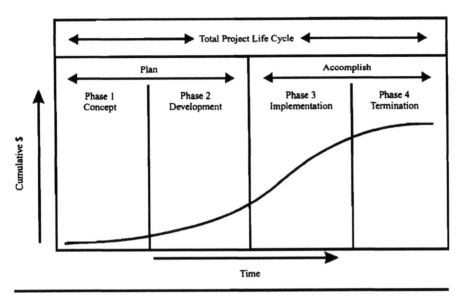

FIGURE 3.5 Cumulative cost over time curve.

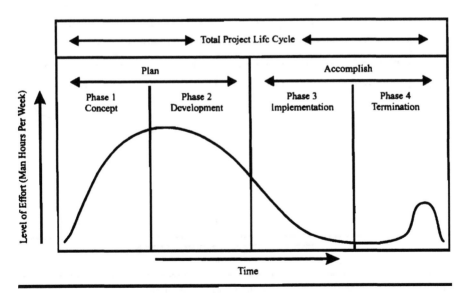

FIGURE 3.6 Project manager's level of effort curve.

project completion reports, negotiating with regulators, and ensuring that "as-built" documentation is provided. At this point in the project, much of the project team will have been released.

3.8.2.4 Project Manager's Cost Over Time Curve

The project manager's cumulative cost over time curve (Figure 3.7) reflects the level of effort curve. The cumulative cost over time curve for the project manager is very steep in the initial phases and less steep in the implementation phase. Finally, this curve becomes steep at the very end of the project. Note that this is in contrast to the cumulative curve for the cost of the project as a whole, which is very steep in the implementation phase.

3.8.2.5 Potential for Reducing Costs

The possibility of identifying a strategy that has the potential for reducing overall costs is greatest during the first two phases of a project, conception and development (see Figure 3.8). This is because during the early phases:

- Alternatives are evaluated.
- Economic analysis can be performed.

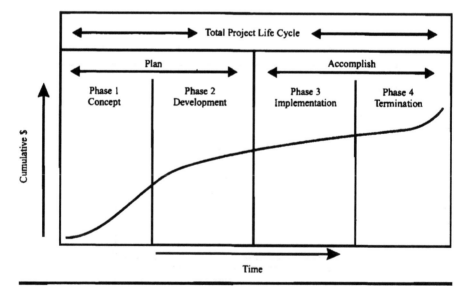

FIGURE 3.7 Project manager's cumulative cost over time curve.

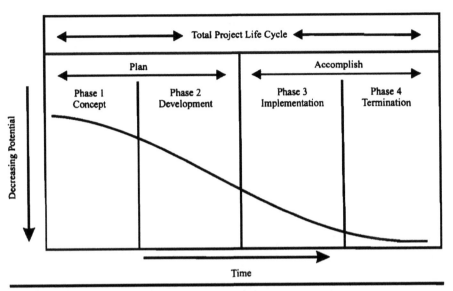

FIGURE 3.8   Potential for reducing costs over time. (From Wideman, R.M., *A Framework for Project and Program Management Integration,* Project Management Institute, Drexel Hill, PA, 1991, iii-7. Reprinted from *A Guide to the Project Management Body of Knowledge* with permission of the Project Management Institute, 130 South State Road, Upper Darby, PA 19082, a worldwide organization of advancing the state-of-the-art in project management. Phone: (610) 734-3330, fax: (610) 734-3266.)

- The least amount of labor hours per unit of time is required (see level of effort curve in Figure 3.4).
- The least amount of money is invested.

Even though the potential for reducing costs is greatest at the beginning, many customers and project managers for that matter often choose to forego planning, preferring instead an attitude of "Let's just go do it." This is because many people do not view planning as work or accomplishing anything and there is a strong desire to see something tangible toward project completion.

However, Table 3.1 has been inserted as an example of the amount of money that can be saved through planning. From a pure project management standpoint, feasibility studies (as required by CERCLA) and corrective measure studies as required by the RCRA can be considered as stages of the planning phases of a project; these studies can be seen as government-mandated planning. Most customers who face the prospect of such a study would not interpret this as good news. Yet the table indicates substantial savings associated with the accepted alternative after completion of the study over what was being discussed by the

TABLE 3.1   Cost Savings from Planning

Facility type	Active railyard	Defense Dept. contractor	Manufacturer plant
Regulatory driver	CERCLA	RCRA	State program
Study type	Feasibility study	Corrective measures study	Evaluation of alternatives
Cost of study	$950,000	$80,000	$10,418
Number of technologies included in screening[a]	39	35	26
Technologies remaining[a] after screening/treatability studies	27	5	7
Number of alternatives[a]	14	2	4
Cost range of alternatives[a]	$550,000– 44,360,000	$1,152,000– 3,191,000	$230,000– 1,704,000
Cost of recommended alternative[a]	$8,410,000	$1,152,000	$301,000
Estimated cost savings	$15,800,000	$2,000,000	$300,000
$ saved per dollar invested	$17	$25	$29

[a] Soil only.

regulatory agency prior to completion of the study. In the CERCLA case (see Table 3.1), a savings of over $15 million was realized. From the data provided in the table, an average of $23 saved in implementation cost for each dollar spent on the studies performed can be calculated. The information presented in this table is from actual environmental remediation projects.

3.8.2.6   *Escalating Cost to Fix or Change*

The cost of making changes is lowest in the first two phases of a project.[22] The cost to change the scope of work then rises more steeply as the project progresses (see Figure 3.9); an example is the construction industry, where it has been said that the cost to make a change increases ten times in each succeeding phase.[22]

3.8.2.7   *Quality of Information*

Unfortunately, the quality of information is at its lowest in the initial phases of a project. This means that planning must be accomplished using limited infor-

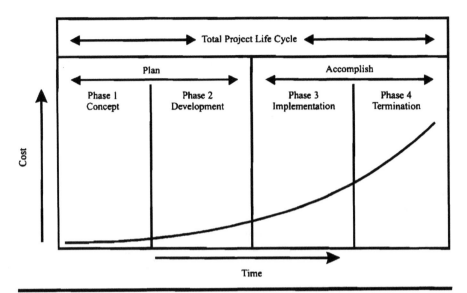

FIGURE 3.9 Escalating cost to fix or change over time. (From Wideman, R.M., *A Framework for Project and Program Management Integration*, Project Management Institute, Drexel Hill, PA, 1991, iii-8. Reprinted from *A Guide to the Project Management Body of Knowledge* with permission of the Project Management Institute, 130 South State Road, Upper Darby, PA 19082, a worldwide organization of advancing the state-of-the-art in project management. Phone: (610) 734-3330, fax: (610) 734-3266.)

mation or with information that is not known to have a great deal of accuracy. As a project progresses, more information is gathered. During the implementation stage, for instance, a great deal is discovered regarding the quality of planning and construction designs. Anyone who has ever worked on an environmental remediation construction project knows that a great deal of information is gathered about the extent of impact at a site during the implementation phase. The curve for the quality of information over time is depicted in Figure 3.10.

3.8.2.8   *Risk versus Amount at Stake*

As stated in the previous section, the quality of information is lowest in the initial stages of a project. The lower the quality of information, the greater the uncertainty associated with the project. Uncertainty is always directly correlated with risk. For example, if a person betting on a dice game knew what the next number was going to be, there would be no risk associated with the bet. Since the number is not known, however, the person's money is at risk. Because the

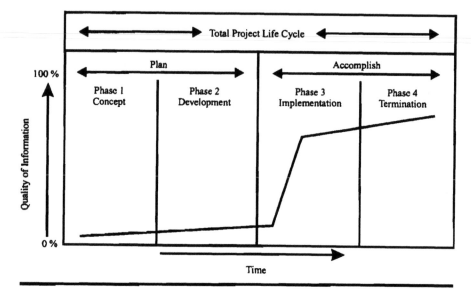

FIGURE 3.10   Quality of information over time. (From Wideman, R.M., *A Framework for Project and Program Management Integration,* Project Management Institute, Drexel Hill, PA, 1991, iii-9. Reprinted from *A Guide to the Project Management Body of Knowledge* with permission of the Project Management Institute, 130 South State Road, Upper Darby, PA 19082, a worldwide organization of advancing the state-of-the-art in project management. Phone: (610) 734-3330, fax: (610) 734-3266.)

quality of information is lowest in the initial phases of a project, uncertainty and risk are high. Uncertainty does not start to fall significantly until implementation, when the unknowns progressively translate into knowns.[23]

There is an interesting correlation between the amount at stake and risk. The amount at stake, measured in terms of resources invested (money spent), is relatively low during the first two phases of a project — but it rises rapidly during the implementation phase. Note that even though uncertainty and risk are high in the initial phases of a project, the amount invested at this point in time is relatively low. Figure 3.11 presents a correlation between risk and amount at stake over the course of the project life cycle.

As indicated in Figure 3.11, the period of highest risk impact occurs during implementation. This is the point where the accuracy of our planning is discovered. Here is where we discover if we developed an adequate budget, if our scheduled time is sufficient, if our characterization of the site is accurate, and if our remedial technology is going to perform as planned. Therefore, even though risk in qualitative terms is highest during the initial phases, the financial impact of this risk may be revealed in later phases.

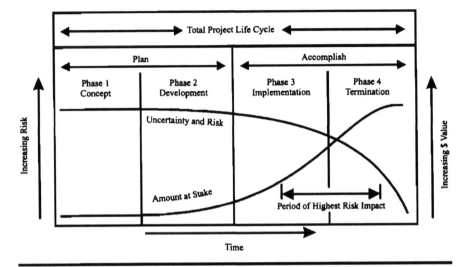

FIGURE 3.11   Risk and amount at stake over time. (From Wideman, R.M., *A Framework for Project and Program Management Integration*, Project Management Institute, Drexel Hill, PA, 1991, iii-9. Reprinted from *A Guide to the Project Management Body of Knowledge* with permission of the Project Management Institute, 130 South State Road, Upper Darby, PA 19082, a worldwide organization of advancing the state-of-the-art in project management. Phone: (610) 734-3330, fax: (610) 734-3266.)

3.8.2.9   Project Stakeholders and the Life Cycle

Project stakeholders are interested parties and participants who have a vested "stake" in the outcome of a project. In the concept phase of a project, the project manager must work to identify the project stakeholders and their objectives. In the development phase, plans are formulated regarding the management of these stakeholders. In the implementation phase, these plans are followed to ensure that the stakeholders do not negatively impact project outcome. A detailed discussion of project stakeholder management is provided in Section 4.11 of this book.

3.9   THE ENVIRONMENTAL INDUSTRY LIFE CYCLE

We can now move on to a discussion of the project life cycle as it applies in particular to the environmental remediation industry. This discussion is focused on the macro level of phases and stages (industry specific) rather than the micro level of tasks (project specific). The work breakdown structure discussed in Chapter 6 provides a more in-depth look at project-specific tasks. The project life cycle as applied to the environmental remediation industry, indicating the general stages associated with the various phases, is presented in Figure 3.12.

Stages such as background data reviews, development of initial investigative work plans, execution of initial investigations, development of extended investigation work plans, and performance of extended investigations constitute the concept phase. These stages will aid in understanding the degree of the problem at hand.

Once it has been determined that the site has been adequately characterized, it is possible to move into the next phase, **development.** This second phase further involves our understanding of the site and ends in the completion of the remedial technology design. During development, the stages may include a risk assessment, feasibility studies or some other form of technology screening, negotiations regarding the technology to be implemented, and the design of the remedial technology. It should be noted that, depending on the results of the risk assessment or the feasibility study as well as regulatory negotiations, the project may end in this phase, due to the selection of the no further action alternative.

It may seem odd to consider risk assessment and a feasibility study as planning, since they represent such a significant amount of effort. However, if one steps back to look at the entire process, it can be seen that these stages are, in fact, a planning mechanism to determine the best response to the problem of impacted soil and groundwater.

Phase three in the project life cycle, **implementation,** would involve the installation, operation, monitoring, and maintenance of the remedial technology. Note that the completion of all these activities will get us to the point of achieving cleanup.

Finally, we come to phase four, **termination.** This involves the development of closure reports as well as negotiating closure with the appropriate regulatory agencies.

3.9.1   Environmental Programs and Life Cycle Stages

The names of the life cycle stages for various environmental regulatory programs are presented Table 3.2. The table presents the goals and terminology for the stages associated with the various regulatory programs (RCRA, CERCLA, Department of Energy, etc.).

3.9.2   Foresight

An important point to bear in mind regarding the environmental remediation project life cycle is that there are other stages and phases to be performed in addition to the one being worked on at any given time. Many project managers

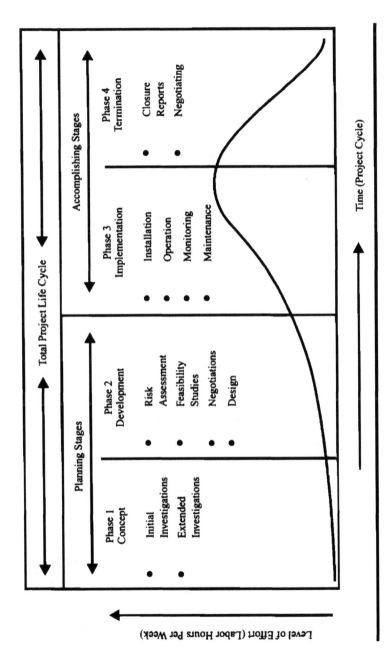

FIGURE 3.12 Environmental project management life cycle.

TABLE 3.2 Life Cycle Stages for Various Environmental Programs

Project life cycle	Environmental remediation goal	RCRA	CERCLA	USTs	DOE	DOD
Phase 1: Concept	Identify releases needing further investigation	RCRA Facility Assessment (RFA)	Preliminary Assessment/Site Inspection (PA/SI)	Phase 1: Preliminary site evaluation	Development phase	Assessment and characterization
	Characterize nature, extent, and rate of release	RCRA Facility Investigation (RFI)	Remedial Investigation (RI)	Phase 2: Site characterization	Implement remedy	Environmental review
Phase 2: Development	Evaluate alternatives and select remedy	Corrective Measures Study (CMS)	Feasibility Study (FS)	Phase 3: Additional site characterization and evaluation of site remedies	Implement remedy	Feasibility Study (FS)
		Public involvement	Public involvement		Title 1	Selection of DOD alternative
	Design remedy	Corrective Action Decision (CAD)	Record of Decision (ROD)	Phase 4: Remedial Action Plan (RAP)	Title 2	Engineering
			Remedial Design (RD)			
Phase 3: Implementation	Implement remedy	Corrective Measure Implementation (CMI)	Remedial Action (RA)	Execute Remedial Action Plan (RAP)	Construction/ Title 3	DOD operations
Phase 4: Termination	Maintenance and surveillance of site	Postclosure monitoring	Operations & Maintenance (O&M)	Operations & Maintenance (O&M)		Closeout
		Verification	Verification	Site closure plan		Verification

Material reproduced by permission of S.M. Stoller Corporation, 5700 Flatiron Parkway, Boulder, CO 80301. Phone: (303) 546-4448, fax: (303) 440-6287.

within the environmental industry have neglected this fact. This may cause the project manager to overlook some elements in the earlier stages that would facilitate the work at later stages.

For example, if developing a work plan for an extended investigation on a project that is likely to involve a risk assessment, it would be prudent to collect additional samples and data that might be required to support such an assessment. The last thing a customer wants to hear when beginning a new stage such as risk assessment is that more samples and additional field activities must be performed in order to gather the necessary data.

In a similar fashion, if it is expected that one of the bases to be offered to regulators in support of closure is the percentage of mass removal, it would be important to have actually determined the mass of contaminant in place during the investigation stage and to have monitored the total mass removed during the operations stage.

3.10   THE PROJECT PLANNING AND CONTROL CYCLE

As we have seen, planning implies the ability to anticipate and prepare actions that will bring about a desired outcome. Planning is an important first step in project completion.

The project manager is key to successful project planning, and it is desirable for him or her to be involved from conception to completion. Such involvement provides continuity and commitment.

All projects are unique events that utilize a limited amount of resources. In addition, multiple projects are in competition for the same resources. In order for these resources to be distributed adequately, formal detailed planning is necessary — on all projects.

It should be noted that planning environmental remediation is typically done at the proposal stage under tight time constraints (the bid due date). The proposal submitted in many instances represents the project plan. Because of the tight time constraints, there can be a tendency to pull a proposal together quickly and not worry about planning until the project is awarded. This type of approach almost always guarantees project failure.

The project planning and control cycle, as presented in Figure 3.13, can be used as a standardized process to facilitate planning under tight time constraints. The boxes in the upper half of this figure represent the planning activities and the boxes in the lower half represent tracking or monitoring of the planned activities. Note that the project budget (as indicated by Figure 3.5, cumulative cost over time) should fully support these planning and control activities.

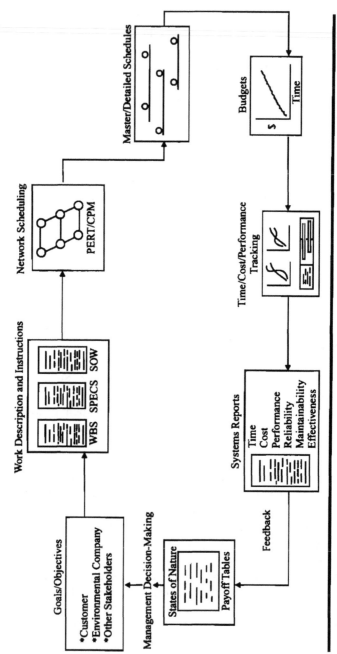

FIGURE 3.13 The project planning and control cycle. (From Kerzner, H., *Project Management: A Systems Approach to Planning, Scheduling and Controlling,* 3rd ed., Van Nostrand Reinhold, New York, 1984, 577. Reprinted by permission of John Wiley & Sons, Inc.)

The sequence of Chapters 4 through 9 of this book closely matches the sequential planning process as presented in Figure 3.13 (up through the block titled time/cost/performance tracking). Each of these chapters provides tools, such as forms, checklists, and rules of thumb, to facilitate the planning process. Additionally, the items identified in Figure 3.13, such as the work breakdown structure (WBS), statement of work (SOW), program evaluation and review technique (PERT), and critical path method (CPM), are discussed in detail in Chapters 6 and 7.

3.11 PLANNING DEFINITION

Project planning can be defined as the process of developing, in detail, all of the project criteria (as presented in Section 3.5) in accordance with the specific sequence presented in Figure 3.13. The planning is performed to the point that all work to be accomplished is readily identifiable to the project team members and the customer.

For a project manager, there are five basic reasons for planning:[24]

- To obtain better understanding of the objectives
- To eliminate or reduce uncertainty
- To improve efficiency of the operation
- To define all work required such that it is readily identifiable to the customer and project participants
- To provide a basis for monitoring work

It should be noted that planning is often an iterative process, as indicated by Figure 3.13. As project tracking is performed, and more is learned about the project as it proceeds along the life cycle, the scope of work may be altered — and that means plans need to change. In many instances, this calls for yet another trip through the planning cycle.

One way to think about project planning is as being similar to designing a chemical process system. Looking closer at a network diagram involved in project scheduling (Chapter 7), an analogy can be seen between:

- The network diagram
- The engineering process flow diagram for a remediation system

In essence, by placing the project data into a project management software package, a computer model or simulation is being developed. This is no different from the computer models used to simulate the fate and transport of groundwater contaminants in hydrogeological systems. Obviously, the more we compare our models with actual results, the more accurate we can make our future models. Perhaps if planning were thought of as designing a process system, individuals would find designing such a system enjoyable and would be challenged to develop effective detailed plans.

3.12 EXERCISE

As stated in Section 1.6, this book grew out of a three-day training program on project management techniques for environmental remediation professionals. During the course, the attendees would separate into small working groups and work on exercises assigned at the end of a series of one-hour lectures. The lectures outlined the material contained in Chapters 4 through 9 of this book. Completing the exercises resulted in a detailed project plan for a former compressor manufacturing site owned by the fictitious Mega Manufacturing Corporation. In addition, these exercises demonstrated how earned value analysis could be used to determine if a project is proceeding in accordance with the plan. The exercises performed by the work groups are included at the end of Chapters 4 through 9.

Appendix A contains summary information for a project known as the Megacorp site. It describes the site and the circumstances surrounding a newly released request for proposal. The exercise for this chapter is to read the summary. The summary contains political, legal, economic, and regulatory information, in addition to technical data. The summary is filled with complexities, uncertainties, and pitfalls, similar to any remediation project. Despite the complexity and uncertainty, the site owner is looking for a service company that will provide a firm fixed price for investigating the site, pilot testing a remedial technology, and designing a remediation system.

Providing a fixed price for pilot testing is rather difficult since the site has not been fully investigated and the remediation service company does not know for certain which technology it would want to pilot test. By the same token, it is difficult to provide a fixed price for design without a completed investigation and pilot test. A request to provide a firm fixed price under these circumstances is in alignment with current trends in the industry. The techniques provided in this book are intended to help environmental project managers effectively deal with such situations.

When reading the remaining chapters, reflect on the Megacorp site information. This will enhance understanding of the material contained in the chapters and prepare you for performing the exercises at the end of each chapter, if you so choose. Answers to select exercises are provided in Appendix F.

REFERENCES

1. Cleland, D.I. *Project Management, Strategic Design and Implementation,* McGraw-Hill, New York 1990, 2
2. Davis, R.C., *The Fundamentals of Top Management,* Harper and Brother Publishers, New York, 1951, 268
3. Kerzner, H., *Project Management: A Systems Approach to Planning, Scheduling, and Controlling,* 2nd ed., Van Nostrand Reinhold, New York, 1984, 2.
4. Newman, W.H., Kirby Warren, E., and McGill, *The Process of Management: Strategy, Action, Results,* 6th ed., Prentice-Hall, Englewood Cliffs, NJ, 1987, 140.
5. Project Management Institute Standards Committee, *Project Management Body of Knowledge (PMBOK),* Project Management Institute, Drexel Hill, PA, 1987, 0-2.
6. Cleland, D.I. and King, W.R., *Project Management Handbook,* 2nd ed., Van Nostrand Reinhold, New York, 1988, 129.
7. Duncan, W.R., Toward a revised PMBOK document, *Project Management Journal,* Vol. XXIV, No. 2, 4, 1993.
8. PMI Standards Committee, *A Guide to the Project Management Body of Knowledge,* Project Management Institute, Upper Darby, PA, 1996, vii.
9. Kerzner, H., *Project Management: A Systems Approach to Planning, Scheduling, and Controlling,* 2nd ed., Van Nostrand Reinhold, New York, 1984, 3.
10. Kerzner, H., *Project Management: A Systems Approach to Planning, Scheduling, and Controlling,* 2nd ed., Van Nostrand Reinhold, New York, 1984, 4.
11. Kerzner, H., *Project Management: A Systems Approach to Planning, Scheduling, and Controlling,* 2nd ed., Van Nostrand Reinhold, New York, 1984, 6.
12. PMI Standards Committee, *A Guide to the Project Management Body of Knowledge,* Project Management Institute, Upper Darby, PA, 1996, 6.
13. Wideman, R.M., *A Framework for Project and Program Management Integration,* The PMBOK Handbook Series, Vol. 1, Project Management Institute, Drexel Hill, PA, 1991, II-4.
14. Project Management Institute Standards Committee, *Project Management Body of Knowledge (PMBOK),* Project Management Institute, Drexel Hill, PA, 1987, 1-2.
15. Wideman, R.M., *A Framework for Project and Program Management Integration,* The PMBOK Handbook Series, Vol. 1, Project Management Institute, Drexel Hill, PA, 1991, II-5.
16. Project Management Institute Standards Committee, *Project Management Body of Knowledge (PMBOK),* Project Management Institute, Drexel Hill, PA, 1987, 1-3.
17. Wideman, R.M., *A Framework for Project and Program Management Integration,* The PMBOK Handbook Series, Vol. 1, Project Management Institute, Drexel Hill, PA, 1991, II-6.
18. Project Management Institute Standards Committee, *Project Management Body of Knowledge (PMBOK),* Project Management Institute, Drexel Hill, PA, 1987, 2-2.
19. Wideman, R.M., *A Framework for Project and Program Management Integration,* The PMBOK Handbook Series, Vol. 1, Project Management Institute, Drexel Hill, PA, 1991, IV-2.
20. Wideman, R.M., *A Framework for Project and Program Management Integration,* The PMBOK Handbook Series, Vol. 1, Project Management Institute, Drexel Hill, PA, 1991, III-4.

21. Wideman, R.M., *A Framework for Project and Program Management Integration,* The PMBOK Handbook Series, Vol. 1, Project Management Institute, Drexel Hill, PA, 1991, III-5.
22. Wideman, R.M., *A Framework for Project and Program Management Integration,* The PMBOK Handbook Series, Vol. 1, Project Management Institute, Drexel Hill, PA, 1991, III-6.
23. Wideman, R.M., *A Framework for Project and Program Management Integration,* The PMBOK Handbook Series, Vol. 1, Project Management Institute, Drexel Hill, PA, 1991, III-8.
24. Kerzner, H., *Project Management: A Systems Approach to Planning, Scheduling, and Controlling,* 2nd ed., Van Nostrand Reinhold, New York, 1984, 534.

Reviewing Requests for Proposal

4

Although this chapter is written from the standpoint of project managers who work within an environmental service company — who must create detailed project plans to service their customers' needs — all can benefit from it. Important topics covered in this chapter include:

- Summarizing complex data
- Developing goals and objectives
- Dealing with political issues
- Performing stakeholder analysis
- Evaluating project risks

These topics are presented within the context of responding to a request for proposal, because often in the environmental remediation industry, the scope of work presented in a submitted proposal acts as the work plan for the project at hand. This means that quality planning must begin at the proposal stage of a project.

4.1 RELATIONSHIP OF THE PROPOSAL TO THE PROJECT PLANNING AND CONTROL CYCLE

From the perspective of an environmental service company, the receipt of a request for proposal (RFP) — and the subsequent decision to respond to the request — represents the beginning of the project planning and control cycle (see Section 3.10 and Figure 3.13). From the site owner's perspective, the project planning and control cycle begins with the internal decision to develop and issue the RFP.

4.2 IMPORTANCE OF A HIGH-QUALITY PROPOSAL

A proposal is the major tool by which a service company secures new business; therefore, the importance of a proposal cannot be overemphasized. The ability to attract and successfully respond to RFPs is a strong indicator of the long-term potential of any service company. A successful proposal meets two primary criteria:

- The proposal is approved by the customer's review process.
- The proposal is executable as specified.

Nothing is gained other than trouble and complaints if a convincing proposal is submitted and awarded, but the proposing company cannot come through.[1] The typical response time allotted in most RFPs is two weeks. This is a relatively short amount of time when one considers the significant amount of work that must be accomplished.

Within the RFP response time:

- The information in the RFP must be fully understood.
- A strategic approach must be developed and estimated (schedule and cost).
- The approach must be clearly articulated.
- A sales theme must be woven throughout the text of the proposal.

4.3 A PROPOSAL IS A PROJECT

Developing a proposal is a project in itself. This idea can clearly be seen by revisiting the definition of a project (see Section 3.4.1). In the following discussion, the words in bold type mirror the defining elements of a project.

Developing a proposal involves the completion of a series of **tasks** which have as their **objective** the submission of a successful proposal. There is a defined **scope of work** for the completion of the proposal, which involves determining the technical approach, developing a work breakdown structure, designing a network schedule, estimating costs, and the actual writing of the proposal. These tasks must occur within a **schedule for completion** as defined by the RFP. There is a definite **cost** for preparing a proposal, such as the base salary of the personnel working on the proposal and the cost of the materials, equipment, and subcontractors. Although many companies neglect to consider the cost of proposal preparation, prudence requires developing a **budget** for proposal preparation and tracking preparation costs. Finally, **resources** are required for completion of a proposal, many of which are limited.

4.4 KEEPING PROPOSAL COSTS IN LINE

A general rule of thumb is that the cost to prepare a proposal should not be more than 10% of the estimated gross cost of a project (cost refers to the dollar amount that will be incurred by the service company in completing the project, not the price the customer will be charged). One author has indicated by means of a computer modeling technique that net profit is maximized when the cost to produce a winning proposal is approximately 7% of the estimated gross cost of the project.[2]

4.5 RECOMMENDED PROPOSAL FORMAT

The proposal format should reflect both project criteria (see Section 3.5) and the project planning and control cycle (see Section 3.10). The recommended proposal format is as follows:

1.0 Introduction
2.0 Background information
3.0 Objectives
4.0 Technical approach
5.0 Scope of work
6.0 Project team and related experience
7.0 Schedule
8.0 Costs
9.0 Assumptions and terms and conditions

The **introduction** is used to briefly describe the events that led up to the submission of the proposal, as well as to introduce the service company.

The **background information** provides the opportunity for the service company to demonstrate its understanding of the project. Additionally, this section should call out any gaps in data that will need to be addressed by the detailed scope of work, enabling cost-effective completion of the project.

The **objectives** section is used to restate the objectives as identified by the customer and, more importantly, to state the short- and long-term objectives for the project as identified by the remediation service company.

The **technical approach** presents the basic strategy that has been developed in order to meet the objectives.

The **scope of work** provides a discussion of the details of how the technical approach will be implemented.

The **project team** section provides the remediation service company with the opportunity to introduce the selected personnel and their experience on related projects.

The **schedule** section of the proposal should present the duration of the project and various tasks, while showing how these tasks are linked and interdependent. See Chapter 7 for detailed coverage of developing network schedules.

The **cost** section of the proposal should illustrate the cost of the various tasks, the cost of the project as a whole, and the cumulative cost over time curve, for subsequent earned value analysis. Chapter 9 provides a detailed discussion of earned value analysis.

Assumptions and **terms and conditions** provide the boundaries for the project scope of work. This section tightens up the details of the scope of work and, if well written, protects both the remediation service company and the customer. Note that the assumptions are typically developed and recorded throughout the planning of the project, whereas the terms and conditions are often based on company policy.

In this section, care should be taken to inform the customer that the assumptions and terms and conditions are based on the project scope. If new information, gained during execution of the project, indicates that an assumption was invalid or a change in terms and conditions is required, the project scope of work will be affected, thereby making a change order necessary.

4.6 BASIS FOR THE BID–NO BID DECISION

Upon receipt, all RFPs must be reviewed for the purpose of:

- Identifying data gaps in the RFP information
- Identifying the customer's objectives (as well as those of other stakeholders)
- Identifying the customer's special requirements
- Determining whether or not the project is worth bidding on

The above information provides the basis for the bid–no bid decision as well as the scope of work to be included in the proposal response.

4.6.1 The Bid–No Bid Evaluation Process

A three-step process is recommended for bid–no bid evaluations on all larger projects. This process can also be applied to smaller projects, if desired by a service company's operations management.

The **preliminary bid decision** should be performed by operations management, assisted by sales and marketing personnel. Operations management will indicate the completion of this task by filling in Sections A, B, and C of the Bid–No Bid Evaluation Form provided in Appendix B.

A preliminary decision by operations management to bid on a project indicates that the project is

- In line with the company's business policy and strategy
- In the realm of the company's capabilities, considering technical requirements and staff availability
- Likely to yield an acceptable profit

The **intermediate bid decision** is based on a review of the terms and conditions of the contract as provided in the RFP. Analysis of the terms and conditions should be performed by a contracts administration group, which will then fill in Section D of the Bid–No Bid Evaluation Form. Operations management and the contract administration group will review the information contained in Section D of the Bid–No Bid Evaluation Form, and the intermediate decision will be made.

The **final bid–no bid decision** will include input from a number of individuals, including the proposal manager and his or her support team. Note that while the proposal manager may or may not be the project manager, should the project be awarded it is recommended that the project manager be involved in the project from the start (see Section 3.10). It is prudent, therefore, to have a proposal manager who will ultimately act as project manager if resource constraints permit. The final bid–no bid decision requires completion of Sections E to H of the Bid–No Bid Evaluation Form. The information contained in Sections E to H should be reviewed by operations management and the final bid–no bid decision then made.

4.6.2 The Proposal Manager and Team

It is during the intermediate decision that a proposal manager should be assigned. Additionally, a proposal review team should be assembled to assist in a detailed review of the proposal. Note that if a proposal review team has not been assigned, the proposal manager should make a request for selected team members — based on a preliminary review of the RFP.

In most cases, a proposal review team will only be required on large-scale industrial (RCRA/CERCLA) projects. This is because RFPs for RCRA/CERCLA projects are often accompanied by a number of voluminous reports that contain historical and environmental investigative data. A typical proposal review team should include (in addition to the proposal manager):

- A geologist (mid-level professional)
- An engineer (mid-level professional)
- A sales representative
- A project administrator

The proposal review team must analyze all available information and summarize it for other decision makers, thereby facilitating the final bid–no bid decision.

A further benefit of the review team's detailed analysis and summary is that the initial technical approach and sales themes may begin to become apparent through the completion of the RFP Summary Form and the Stakeholder Management Summary Form (see Sections 4.2.5 and 4.6.7).

Development of the final technical approach and sales themes, in most cases, will require input from the key project personnel. The key project personnel are discussed in detail in Section 5.1. In some cases, the key project personnel will be the same as the proposal review team. In other cases, the key personnel will consist of more experienced, senior technical specialists.

4.6.3 The Proposal Number

An account number, similar to a project account number, should be established for tracking all proposal costs. This number should be established at the time the proposal manager is assigned, even though the final decision to bid on a project has not been made.

The reason for assigning a proposal number at this point is twofold. First, there is a high probability that the final decision will be in favor of bidding on the project, since the preliminary and intermediate bid decisions were positive. Second, it is likely that the detailed review and analysis of the proposal could involve a significant cost (two to three days worth of time). Duration of this phase and the associated costs will depend on factors such as the amount of information provided in the RFP and/or the need for a site visit or prebid meeting.

4.7 SUMMARIZING DATA

Summarizing data is a crucial step in responding to an RFP. RFPs for environmental remediation projects can be specific and contain a vast amount of information. On the other hand, some customers issue RFPs that are quite vague, with very little useful information provided.

4.7.1 A Vague RFP

Caution should be exercised when bidding on a project for which a vague RFP has been supplied. A vague RFP usually indicates one of two scenarios. The first — and more benign — is that the customer simply does not know how to develop an appropriate RFP. The second scenario — which is worse — is that the customer is searching for an approach; upon receipt of all proposals, the customer will reissue the most promising approach in a separate RFP. Note that in the second scenario, the customer usually has no intention of awarding the original RFP. Thus, if the second scenario appears to be the case, it usually is best to not bid on such a project since it will involve only a waste of time and money.

4.7.2 Sources of RFP Data

Regardless of the amount of information provided in the RFP, all data regarding the site must be summarized. Available data can come from:

- The sales representative who brought in the proposal
- Phone calls with the customer's designated representative(s)
- A prebid site visit

It should be noted that if a prebid site visit has not been offered by the customer, a request should be made to visit the site. Bidding on a project without visiting the site introduces unnecessary risk.

4.7.3 Categories of Information

One of the best ways to summarize the information associated with a given project is to organize the information in the form of bulleted statements under a number of category headings, thereby creating an outline. In this manner, one can obtain a better understanding of the site and begin to identify gaps that may be present in the information. (As mentioned previously, the scope of work as outlined in the submitted proposal will often involve activities designed to resolve these data gaps.)

Information can be broken into subcategories and subsubcategories. The number and types of categories associated with each project may vary depending on the project. The typical information categories associated with an environmental remediation project include:

- Background information
- Physical characterization

- Analytical characterization
- Regulatory requirements

An example of subcategories pertains to background information, which can be further subdivided into plant operational history and current site conditions. Another example is physical characterization, which can be broken down into regional geology and site geology.

It is recommended that the information in each of the subcategories be grouped under two headings: known information and unknown information.

4.7.4 RFP Summary Form

The RFP Summary Form in Appendix B is provided to assist the core team in assembling information. An outline of the RFP Summary Form is presented below:

1.0 Background information
    1.1 Operational history
    1.2 Current site conditions
    1.3 Previous environmental activities (if any)
2.0 Physical characterization
    2.1 Regional geology
    2.2 Site geology
3.0 Analytical characterization
    3.1 Soil analytical results
    3.2 Groundwater analytical results
    3.3 Sediment analytical results
4.0 Regulatory issues
5.0 Other

A detailed discussion regarding the types of information that should be placed within each category is unnecessary. It would be quite lengthy and in most cases the information is self-evident based on a review of the category headings. Furthermore, such a discussion could serve to limit the creativity of the individuals performing a summary. Nevertheless, a couple of brief examples are provided.

4.7.4.1 Sample Background Information: Operational History

The known information for operational history might include:

- Length of time at location
- Type of business(s)

- Chemicals used in manufacturing processes
- Historical waste-handling procedures

The unknown information under this category might include:

- Dates of changes in manufacturing processes
- Inventory of all chemicals historically used at the site
- Time period of any releases of chemicals into the environment

4.7.4.2 Sample Physical Characterization: Site Geology

The known information regarding site geology might include:

- Soil type(s)
- Depth to bedrock
- Approximate groundwater elevation

The unknown information in this category might include:

- Groundwater direction of flow and gradient
- Hydraulic conductivity of saturated soil
- Background concentration of metals in soil

4.7.5 Site Visit/Data Gathering

The purpose of visiting any site is to gather as much information as possible that might influence the final evaluation of a project. On many occasions, site visits are often offered by customers to bidding contractors in the form of a prebid meeting. As previously stated, if the customer has not planned a prebid meeting, the remediation service company should request a visit to the site. The remediation service company must be sure to properly prepare for a site visit by thoroughly reviewing all bid documents including contracts and specifications, so that site conditions can be related to the specific contractual requirements. If allowed, a narrated videotape of the site should be made, to be copied and sent to team members who may be located in other offices. Photographs may also be taken.

To assist remediation service company personnel in performing a site visit, a prebid site investigation checklist is included in Appendix C. This checklist originally appeared in *Skills and Knowledge of Cost Engineering*,[3] a publication by the American Association of Cost Engineers (reprinted with permission). Although this checklist was designed primarily for construction projects, it includes a considerable amount of information that is applicable to environmental remediation projects as well. This tool is especially applicable now that more and more environmental remediation projects are moving into the construction implementation phase.

4.8 IDENTIFYING OBJECTIVES

As stressed in Chapter 3, objectives are fundamental to the definition of a project and therefore must be identified early on.

4.8.1 Mission, Objective, and Goal

In everyday language, the terms mission, objective, and goal are used interchangeably. If, however, we seek a common language for the project management profession, to facilitate project communications, individual definition of these terms is appropriate. As with the definition of a project, these definitions may initially seem self-evident. In project management literature, however, Cleland has provided a clarification of the terms mission, objective, and goal. As defined by Cleland, there is a hierarchy associated with these terms, with mission being the largest and most widely encompassing and goal the smallest.

A **mission** is defined as "a broad enduring intent pursued by an organization."[4] Cleland further defines a mission as "the overall strategic purpose toward which all company resources are directed and committed."[4] The term mission, therefore, seems to be more aligned with an organization's *reason for being* or *function in the world.* For example, a remediation service company may establish as its mission to be a top provider of remediation services.

Projects assist an organization in achieving its overall mission. The use of the term mission therefore does not seem appropriate in the context of a discrete project. Some authors have written about developing mission statements for projects; however, this practice is not widely accepted and seems of questionable value. Mission, then, can be understood to generally relate to the organization.

An **objective** is defined by Cleland as "the target."[4] It specifies the "critical results that must be achieved, in the long term to contribute directly toward the organization's mission."[4] The term objective, therefore, is appropriate in the context of a project.

A **goal** is defined by Cleland as a time-sensitive milestone, the attainment of which signifies that progress has been made toward the project objective.[4] Thus the term goal relates to the context of an objective.

4.8.2 Single or Multiple Objectives

According to Schuyler, there are at least two schools of thought on planning with regard to objectives: the multiple and the single objective.[5] Many project managers believe that it is necessary to list a number of "major objectives" that will have been accomplished upon completion of a project. The single objective school involves stating one single objective, with other considerations identified as subobjectives.

4.8.3 Long- and Short-Term Objectives

Identification of a single long-term objective is recommended for project planning. This **long-term** objective is defined as the desired situation or product that will be in place once the entire project life cycle has been completed. The long-term objective can then be broken down into multiple short-term objectives. These **short-term** objectives are the desired situations or products that will be in place once the life cycle stages have been completed. Short-term objectives can be broken into **subobjectives**, which represent what will be in place at the completion of project tasks.

Finally, in project planning, **goals** are the milestones (steps along the way) leading to the completion of the various objectives. As will be seen in Chapter 6, this hierarchy of objectives is consistent with the development of the project work breakdown structure.

4.8.4 Application to Environmental Remediation

Having reviewed the definitions of objectives (long-term, short-term, and subobjectives) and goals, we can now see how this theory applies to the business of environmental remediation. Identification of objectives for an environmental remediation project (or any project) begins with knowledge of the customer. It can be safely said that the main business of customers of remediation services is something other than environmental cleanup; they are in business to provide a product or service while making a profit for their company and its shareholders.

The performance of an environmental remediation project can only serve to reduce the overall profitability of the customers of remediation service companies. Therefore, one can be assured that one of the internal objectives of the customer (whether stated or not) is to keep environmental costs to a minimum.

To better understand each customer's project, it is important to identify the driving forces. A **driving force** can be defined as anything that causes a desire for change. Driving forces can be internally or externally generated; in many industries, the driving forces are primarily internally generated, such as the desire to offer a new product or service to the market.

In the environmental remediation industry, the primary driving forces are externally generated (e.g., due to regulations or the discovery of a situation that is creating a liability, such as a leaking underground storage tank). However, with the advent of brownfield programs, the driving force for many projects is the desire to return a site to productive and profitable capacity. The remediation project is seen as completed when no more activity is required by the regulatory agencies (regulatory site closure) and/or the liability has been eliminated or greatly reduced.

4.8.4.1 Long-Term Objective for Environmental Remediation

In consideration of the above issues, the long-term objective for nearly all environmental remediation projects is achieving **regulatory site closure**, with **minimum liability**, and at the **lowest possible cost**.

4.8.4.2 Short-Term Objectives for Environmental Remediation

Although the predefined long-term objective can be stated for nearly every environmental remediation project, it will almost never be found stated as such in an RFP issued by the owner or operator of a site. This is because RFPs are typically issued to address one of the project stages, such as a remedial investigation, feasibility study, remedial design, etc. Note that the stated objectives as provided in most RFPs are really short-term objectives for the overall project.

4.8.4.3 Long- and Short-Term Objectives for Marketing

One marketing technique or theme that can be developed throughout a proposal is to provide a statement of the long-term objective (least cost closure), explaining at every reasonable opportunity how the detailed scope of work provided by the remediation service company will achieve the short-term objective(s) stated in the proposal and the long-term objective as well.

4.8.4.4 Subobjectives

As we have discussed, quality planning requires an overall look at a project in terms of:

- The long-term objective
- The short-term objectives, as stated in the RFP
- The technical, regulatory, and political requirements
- Known data gaps

Given this information, we can now break the project down into subobjectives. It is these subobjectives that provide the basis for the proposal response.

The subobjectives will help to directly determine a project's:

- Work breakdown structure
- Detailed scope of work
- Schedule
- Costs

4.8.4.5 Example: The Pragmatic Use of Objectives and Subobjectives

The following illustrates how objectives and subobjectives can help in the proposal response. Let's assume that a potential customer has issued an RFP for the performance of a RCRA Facility Investigation (RFI). Although the customer has stated that its objective is the performance of an RFI, the remediation service company may identify, and discuss in the RFP response, the appropriate long-term **objective** for the site as least cost regulatory closure while retaining minimum liability. As a marketing element, the remediation service company can indicate that the scope of work presented in its proposal was prepared with this foresight in mind.

Through its RFP, the customer has identified the **short-term objective** as the completion of an RFI. In an effort to demonstrate comprehensive planning, the remediation service company could embellish this objective. The embellished objective could be stated as follows:

> This proposal will address the completion of an RFI which will provide sufficient information for the performance of a risk assessment and feasibility study.

Based on this short-term objective, the remediation service company may identify a number of subobjectives for the RFI. These subobjectives could include items such as:

- Regulatory approval of an RFI work plan
- Approximation of the horizontal and vertical extent of soil impact
- Approximation of the horizontal and vertical extent of groundwater impact
- Estimation of the total mass of the compounds of concern in the subsurface environment
- Determination of physical and analytical data required for fate and transport modeling
- Determination of the radius of influence of air sparging/soil vapor extraction technology

4.8.5 Objectives and Stakeholder Analysis

In reviewing the RFP, no discussion of objectives is complete without a discussion of project stakeholders. A stakeholder is any individual or group, either internal or external to the project management team, that shares a stake or interest in the project. Because of their stake in the project, the various stakeholders are likely to react to project decisions in a number of ways, many of which can be detrimental. Therefore tasks can be included in the project scope of work which are specifically designed to address the concerns of these stakeholders. Note that performing a stakeholder analysis is crucial in identifying the project objectives. The concept of stakeholder analysis is presented in Section 4.11.

4.9 SPECIFICATIONS

Specifications figure into a proposal response as well. Specifications address the level of quality at which a product or service will be delivered. In other words, specifications provide a statement of the customer's required standards.

In many engineering and construction RFPs, a series of specifications will be provided by the customer. These specifications can include requirements for design drawings, field equipment, site preparation, installation procedures, etc. Small changes in specifications can cause significant changes in project costs. Therefore, specifications are standards for pricing out a proposal. The effect of specification on cost is easily demonstrated. If a customer were to specify that all monitoring wells are to be cased and screened using #316 stainless steel, instead of PVC casing and screen, this would add considerable cost. Because specifications can have a substantial impact on overall cost, every effort must be made by the project team to identify appropriate specifications inherent in the project. The project planning team must strive to identify such specifications even when they have not been addressed by the customer. Specifications (requirements) may cover:

- Type of quality assurance/quality control procedures utilized in the field regarding analytical samples
- Level of laboratory quality assurance/quality control reporting
- Level of protective equipment used by field personnel
- Acceptable air and water discharge levels permitted by the authorized regulatory agency
- Types of materials that can be utilized in treatment system design due to chemical compatibility
- Type of computer program that will be utilized to store and manage site data

Although the proposal specifications must be complete, note that, due to cost implications, it is important to identify only those specifications that are absolutely necessary to the project. Such specifications would include those regarding professional practices, legal requirements, and the health and safety of site personnel.

4.9.1 Functional versus Detailed Specifications

It should be noted that there are essentially two types of specifications that describe the customer's requirements: functional and detailed. An inappropriate choice between the two can easily permit variance from the customer's expectation, leading to a subsequent failure to meet the requirement.[6] In a **functional specification**, focus is on the functions that the product will provide over its useful period of operation. For example, the functional specification for a monitoring well could be to provide a permanent access point permitting the measurement of any floating hydrocarbons and the collection water from the first encountered saturation zone in the subsurface. A functional specification provides the seller with a wide latitude in providing the product.

A **detailed specification** is very precise in the description of the product. The following is an example of a detailed specification for a monitoring well:

- Drill each monitoring well to a depth of 25 feet using 6.25-inch hollow stem continuous augers.
- Install 10 feet of schedule 40 continuous wrap PVC well screen and 15 feet of schedule 40 PVC well casing.
- Install filter pack to an elevation 2 feet above the well screen.
- Place a 2-foot bentonite pack above the well filter pack and complete the remainder of the well annulus to a depth 3 feet from the surface using cement/bentonite grout.
- Install a 6-foot carbon steel casing with locking cap and cement in place while providing a concrete apron with a minimum diameter of 3 feet and a thickness of 4 inches.

This type of precision in the detailed specification can allow the remediation service company to meet the customer's requirements more easily because the expectations are spelled out in detail.

4.9.2 Recording Specifications

Since specifications will have a significant impact on cost, it is important to indicate — in writing — that the project will be performed in accordance with such specifications. During the proposal review stage, all specifications should be included in the RFP Summary Form (Appendix B). There are two areas in the proposal itself where all specifications can be spelled out:

- The written scope of work
- The assumptions at the end of the proposal

4.10 POLITICAL, ECONOMIC, AND TECHNOLOGICAL CONSIDERATIONS

All proposal decisions and strategies must be made in light of any prevailing political, economical, and technological considerations. These considerations balance one another, whereby none of the three is allowed to take precedence over the others. Figure 4.1 presents these considerations as the sides of an equilateral triangle encompassing the overall project strategy in order to emphasize this point. The overall project strategy (also referred to as the technical approach) is based on a balancing of these considerations.

The project manager is often called upon to make pragmatic decisions within these considerations. For example, a less effective though acceptable technology may be proposed based on the customer's financial situation. When reviewing a proposal, it is important to identify, to the greatest degree possible, all political, economic, and technological considerations, a discussion of which follows.

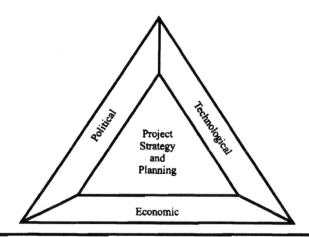

FIGURE 4.1 Balancing considerations in project planning. (From Yaniga, P.M., unpublished data, circa 1989.)

4.10.1   Political Considerations

Most experienced project managers will point out that it is often through the effective use of organization politics that they are successfully able to complete their projects. In fact, the argument can be made that without an understanding of the role that politics plays in project implementation, the likelihood of managing a successful project is greatly diminished.[7] According to Eccles, "politics is about getting collective action from a group of people who may have quite different interests."[8] There are usually three distinct behaviors that are taken by people when they address political considerations. Two of these positions are equally inappropriate, but for entirely different reasons. The third position can help you provide successful project management.

4.10.2   The "Politically Incorrect" Approach

One group of individuals sees politics as something that is unpleasant — and to be avoided at all costs. This rather negative and naive view can be detrimental to project planning, because it does not acknowledge the environment within which a project exists, nor does it seek effective ways of dealing with the political climate.

4.10.3   The "Politically Driven" Approach

The second and exact opposite behavior is taken by individuals who believe that politics and aggressive manipulation are the means to reach their desired goals. This is the group of individuals who are often referred to in the business literature as "sharks." Perhaps the single most distinguishing characteristic of this group is that their loyalty seems entirely toward their own personal objectives — whether or not these objectives will benefit the project, the company, or society at large.[7] Fortunately, the number of individuals who fall into this group is few (although the media might have one suppose otherwise).[7] The extremist positions above underscore the need for project managers who are politically sensible.[7]

4.10.4   The "Politically Sensible" Approach

This means that managers must attempt to execute a project with a pragmatic understanding of how many decisions are made (i.e., through negotiation and bargaining).[14] Such individuals will try to fully understand the political arena in which the project exists and identify strategies accordingly. Understanding the politics of an environment requires one to identify the objectives of the various stakeholders (discussed in Section 4.11). Once the environment is understood, one can be begin to determine those points in the project strategy for which one is willing to negotiate and bargain — and the points where one must hold a firm line. An example of a politically sensible approach would be overdesigning a system for a client who is unwilling to pay for a pilot test that would allow for a more efficient design. Another example of political sensibility would be to refrain from proposing a particular remedial technology — even if it is technically sound — whenever it is well known that the regulators and the public are strongly opposed to that particular technology.

4.10.5   Politics and Stakeholders

Political considerations most often are related to the behavior of a project's external stakeholders. External stakeholders are individuals who have an interest in the project outcome but who are not under the authority of the project manager (e.g., members of the customer's management team).

4.10.6   Economic Considerations

Economic considerations affect all behavior, whether that of a government, corporation, or individual. Usually, economic considerations can be summed up in one simple question: *Is this the best use of available money to address the issue at hand*? Note that there are two elements to this question: the best use of funds and the actual amount available. The best use of funds indicates that there are a number of alternative strategies to solve a problem. The choice among alternatives has led to the development of selection techniques, such as feasibility studies, decision tree analysis, and Monte Carlo computer economic evaluations (see Chapter 10). Regardless of the selection method employed, the point is that we want to identify the best use of available funds.

Many individuals may question why a project manager for a service company would want to seek out strategies that will help the customer save money. Won't this result in reduced fees paid to the service company? The answer is yes in the short term and for the particular project in question. However, a business is built on repeat sales. For example, an adage in the retail business is that unless an individual returns to your store a second time, you do not have a customer. The amount of funding available to a customer for performing a project in particular or in terms of the customer's annual environmental budget is crucial information, since this can indicate the types of project strategies the customer may be willing to support. For example, a decision based on availability of funds would be to install a lower cost, less efficient recovery system because the customer's annual budget cannot support a more efficient/higher capital system.

4.10.7 Technical Considerations

So far, we have covered political and economic considerations for an RFP review. Technical considerations are the least flexible. Technical considerations usually ask the question: *Is it technically possible to achieve the desired objective using this approach?* If in fact it is not technically possible to achieve the desired objectives, obviously the approach must be abandoned. The prudent aspect of technical considerations must be addressed as well. Often, engineers or other technical personnel may seek optimum solutions which go well beyond the customer's needs or desires; many times, such solutions involve performing tasks that are quite costly, given the marginal increased benefit. Therefore, efforts must be made to ensure that this form of going overboard is not included in the project approach.

4.11 STAKEHOLDER ANALYSIS

In project management literature, one of the more recent topics to be discussed is **project stakeholder management**. A **stakeholder** is a group or individual that has (or is believed to have) a vested interest in the outcome of a project. Successful project management requires that responsible managers address the potential influence of project stakeholders. It may not be practical to perform stakeholder analysis on each and every project. Such analysis is most appropriate for large RCRA/CERCLA projects, which are expected to cost several million dollars and last more than three years. However, since the influence of the project stakeholders can (and should) have a direct impact on the objectives identified by the project team, the concept of stakeholder management is discussed here. Additionally, the following information may help you discover new and creative ways to quickly perform stakeholder analysis on smaller projects.

4.11.1 The Project Stakeholder Management Process

The project stakeholder management process consists of the following elements:[9]

- Identifying appropriate stakeholders
- Specifying the nature of their interest
- Measuring their interest
- Predicting what the stakeholders' future behavior might be in order to satisfy their stake
- Evaluating the impact of the stakeholders' behavior on the project team's latitude while managing the project
- Modifying project strategy, in order to manage the stakeholders' influence

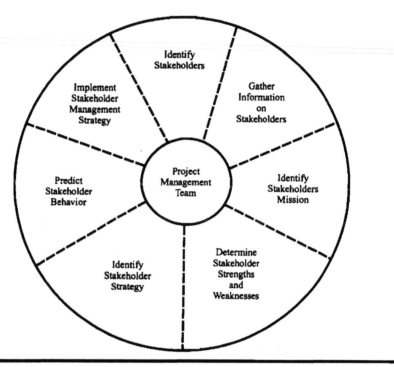

FIGURE 4.2 Project stakeholder management process. (From Cleland, D.I., *Project Management: Strategic Design & Implementation*, 1st ed., McGraw-Hill, Blue Ridge Summit, 1990, 105. Reproduced with permission of The McGraw-Hill Companies.)

Project management author Cleland has further defined the project stakeholder management process as consisting of executing the management functions of planning, organizing, directing, and controlling the resources used to cope with external stakeholder strategies.[10] A model of this management process is presented as Figure 4.2.

4.11.2   Identifying Project Stakeholders

Each project has its own unique set of stakeholders. Stakeholders can be internal to the project (e.g., the project team) or external to the project (e.g., a local community action group). Stakeholders external to the project are sometimes referred to as interveners.

Internal stakeholders are usually (but not always) supportive of project strategies, since they are a part of the team. One way a prudent project manager can ensure the support of internal stakeholders is to include them in the design and development of project strategies, thereby further enhancing their loyalty to said strategies.

External stakeholders, by contrast, may not be supportive of proposed strategies, nor are they under authority of the project manager. Consequently, external stakeholders can provide a formidable challenge in managing.

Examples of external stakeholders for an environmental remediation project include (but are not limited to):

- Government agencies (federal, state, local)
- The current property owner (if different from the party responsible for environmental impacts)
- Current site occupant (lessee)
- The affected local community
- The general public, represented through environmental, social, political, environmental, and other "intervener groups"
- The oversight contractor (if hired by the customer)
- Competitors
- Financial institutions

Figure 4.3 presents a simple model of both internal and external project stakeholders.

4.11.2.1   Influential Stakeholders

Note that stakeholders who attempt to exert an influential role in affecting the outcome of the project should be analyzed and catalogued. In particular, the project manager must address the following major issues:[11]

- Who are the most formidable project stakeholders?
- What are their strengths and weaknesses?
- What is their strategy and the probability of their being able to implement such a strategy?
- Do any of these factors give the stakeholder a distinctly favorable position which can influence the project outcome?

4.11.3   Gathering Information on External Stakeholders

Once the external stakeholders have been identified, the project stakeholder management process calls for gathering information on them (which is similar to gathering information on competitors). In a systematic data-gathering approach, the project manager should ask:[12]

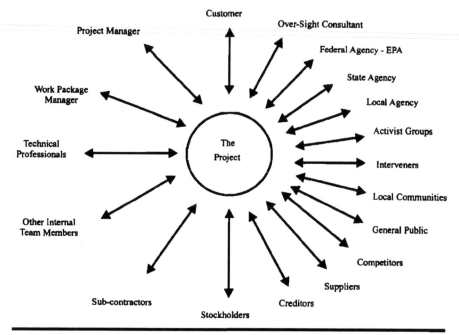

FIGURE 4.3 Project stakeholder network. (From Cleland, D.I., *Project Management: Strategic Design & Implementation,* 1st ed., McGraw-Hill, Blue Ridge Summit, 1990, 105. Reproduced with permission of The McGraw-Hill Companies.)

- What needs to be known about each stakeholder?
- Where and how can this information be obtained?
- What member of the core planning team will gather, analyze, and interpret this information?
- How will this information be used in decision making and strategic planning?

Note that once the stakeholder information has been gathered, it should be provided to experienced project directors or senior-level operations management for analysis. This will enable them to make an informed determination of each stakeholder's target mission.

4.11.4 Identification of Mission

A stakeholder's mission provides the key to understanding the stakeholder's claim on a project, as well as the likely strategy. Sometimes the mission can be found clearly stated in a group's literature. The following is an example of how a stakeholder's mission can affect a project. An organization known as Trout Unlimited became a stakeholder in a particular remediation project. Its mission involved protecting and preserving fishing habitat for the enjoyment of trout fishing. The group became a stakeholder when the EPA issued a ban on trout fishing in a number of streams near a particular project site, due to possible elevated polychlorinated biphenyl (PCB) concentrations in fish.

The loss of these streams resulted in a lawsuit by Trout Unlimited against the identified responsible parties. Once the lawsuit was issued, the strategy, developed by the project team (including the potentially responsible parties and their legal representative), was to collect fish samples for PCB analysis from the streams as well as from local fish markets. This was done in order to perform a statistical analysis as well as a risk assessment, to determine if the EPA had just cause for issuing the fishing ban.

At this point in the project, determining the average concentration of PCBs in fish collected from these streams became one of the subobjectives for the CERCLA Remedial Investigation that the remediation service company was contracted to perform at the site. This example shows how stakeholders can influence the identification of objectives for a project and why stakeholder analysis should be performed.

As a concluding point, it should be noted that not all stakeholder missions are adversarial. In fact, it is a good idea to categorize such missions as either supportive or adverse to the project goals. It is in the best interest of the project manager to keep supportive stakeholders well informed of the project status. On the other hand, potentially adverse stakeholders should be dealt with on a "need to know" basis.[13]

4.11.5   The Adversarial Stakeholder

After a stakeholder's mission has been identified, the next issue is to determine the strengths and weaknesses of the various stakeholders. Knowing these strengths and weaknesses is a prerequisite to understanding the potential success of the stakeholder's strategy.

An adversarial stakeholder's strength and weakness may be based on such factors as:[13]

- Productive use of resources
- Political support
- Public support
- Coherent or incoherent strategy
- Committed, scattered membership

A strategy for dealing with stakeholders can be tested by answering the following questions:[13]

- Does the strategy adequately cope with a strength of the stakeholder?
- Does the strategy take advantage of an adversarial stakeholder's weakness?

4.11.6 Predicting Stakeholder Strategy and Behavior

In order to predict stakeholder strategy and behavior, it is important to determine the *specific* vested interest or stake in the project.

Stakeholders usually perceive a vested interest in a project based on:[13]

- ***Mission relevancy.*** Vested interest directly relates to the cause of the group, as in the example above concerning Trout Unlimited.
- ***Economic interest.*** The stakeholders have an economic interest in the issue. It is highly conceivable that a current landowner would have an economic interest regarding the amount of environmental cleanup to be performed by the previous landowner.
- ***Legal right.*** This interest usually occurs with federal, state, and local regulatory agencies.
- ***Political support.*** These stakeholders see the issue in terms of their political constituency. An example of this is a remediation site where the state governor indicated that one of the goals of his term was to effect proper cleanup of the site.
- ***Health and safety.*** This stakeholder issue is related to the health and safety of a group. A number of lawsuits have been filed by residents who claimed health impacts related to historical operation of hazardous waste sites.
- ***Lifestyle.*** This interest is related to values enjoyed by the stakeholder. Sportsmen and environmental groups often rally around this issue.
- ***Opportunism.*** The interest occurs when the stakeholder involves many others, with the goal of increasing political power at the expense of the project.

Note that once the interest is understood, the expected strategy to be employed by the stakeholder can be identified by researching the methods employed on similar projects. For example, the methods used by local community action groups may be town meetings or other functions designed to generate local media interest. By learning about the historical behavior and strategy of a stakeholder, a project manager is a step ahead in resolving any stakeholder conflicts.

4.11.7 Developing a Stakeholder Management Strategy

Once the most formidable stakeholders have been identified, along with the likely strategy they will pursue to influence the project outcome, a strategy for managing the stakeholders can be formulated. The stakeholder management strategy could include, for instance, developing additional project tasks such as the fish sampling discussed in Section 4.11.4 or conducting off-site residential well sampling. Another method of managing stakeholders is the development of a public relations plan, in conjunction with the customer, in order to ensure that all questions received and information provided are channeled through one central location. It should be noted that if the customer wants the remediation service company to manage stakeholder requests, this could add significant costs to the overall project budget. In such a case, it is recommended that a separate work order be established and that the work be done on a time and materials basis.

In many projects, the most formidable stakeholders are members of the customer's own management team, who have a specific agenda they wish to pursue. Such an agenda could include attempting to replace the remediation service company with a historically favored contractor or a contractor with whom they have a close personal relationship. The Stakeholder Management Summary Form in Appendix B is included in order to help address all the key issues of stakeholder analysis.

4.12 INITIAL RISK IDENTIFICATION

At this point, we are ready to discuss risk. During the review of the RFP, it is essential that the review team identify the associated risks in meeting the project's identified objectives. Risk management is one of the nine project management functions as defined by the Project Management Institute (PMI). It is a large subject area; the basic concepts of risk management and a discussion of how these concepts can be utilized to facilitate the RFP review process are presented in this section. An excellent book on the subject of risk management is *Project and Program Risk Management: A Guide to Managing Project Risks and Opportunities,* edited by R. Max Wideman (available from the PMI).

As discussed in Section 3.8.2.8, risk always directly correlates with uncertainty; the less information available, the higher the uncertainty and the higher the risk. An event about which we are uncertain but which turns out in our favor (such as an investment) is called an opportunity. The possibility of unfavorable results, however, is known as risk.

4.12.1   Definition of Project Risk

Project risk is the cumulative effect of the chances of an uncertain occurrence adversely affecting the project objectives.[15] In other words, project risk is the degree to which project objectives are exposed to negative events and their probable consequences, as expressed in terms of scope, quality, time, and cost.[16]

4.12.2   Project Risk Factors

Project risk is characterized by the following **risk factors**:[17]

- **Risk event** (i.e., precisely what might happen to the detriment of the project)
- **Risk probability** (i.e., the likelihood the event will occur)
- **Amount at stake** (i.e., the extent of loss which could result)

Given the above data, **risk event status** (for any risk event) is calculated by the following relationship:

$$\text{Risk event status} = \text{Risk probability} \times \text{Amount at stake}$$

The **expected value** of all project risks is determined by summing the risk event status for each risk identified. Some companies use the resultant value as the contingency budget when bidding on firm fixed price contracts.

As one might expect, it can be quite difficult to estimate both the risk probability and the amount at stake, should the event actually occur. Numerical methods have been developed by the field of cost engineering to make such estimates using techniques such as decision tree analysis and Monte Carlo computer simulation. These techniques are discussed in detail in Chapter 10.

4.12.3   Conversion of All Risk Impacts to Dollars

Although risk can negatively affect project objectives in terms of scope, quality, and time, as well as cost, for the purpose of risk management, the impact on scope, quality, and time is converted to cost (i.e., a dollar amount). For example, let's assume that during the implementation of a remedial measure involving soil excavation and aboveground treatment, there is a risk of discovering that the extent of impact is twice as large as determined by the remedial investigation. Should such an event actually occur, it certainly would impact the project scope and time of completion; however, this impact can be measured directly in terms of cost. The increased cost would be determined by multiplying the new soil volume times the cost per cubic yard to treat, plus additional administrative costs.

An example of quality risk would be an installed remedial system that does not perform as designed. Quality risk can be measured in terms of dollars by estimating the cost to repair the system in order to meet the quality specifications.

4.12.4   Definition of Risk Management

According to the PMI, project risk management is defined as:

> The art and science of identifying, analyzing, and responding to risk factors throughout the life of a project in the best interest of its objectives.[16]

The systematic approach to these concerns is known as the risk management process.

4.12.5   The Risk Management Process

The risk management process involves four major steps: identify, analyze, respond, and document.[18]

Identification.   One of the best ways to identify project risks is to first break the project down into a number of manageable tasks. This is best accomplished by developing a work breakdown structure as described in Chapter 6. Once the various tasks that have to performed have been identified, brainstorming techniques can be used to consider what might go wrong during the execution of each task. A simple example is hitting an underground obstruction during drilling, getting the rig stuck, or finding that the depth to bedrock is 20 feet beyond the expected depth. All such occurrences will have an impact on the cost of completing a project and, therefore, must be identified.

Analysis.   This step involves assigning a probability and amount at stake to each possible risk event, as discussed in Section 4.12.2.

Response.   The response step involves developing a strategy for how the project team will respond to each risk event should it occur. Often, an appropriate response strategy is not taking responsibility for the risk in the first place. This involves delineating which party (customer or contractor) is the source of the particular risk and therefore best able to control the events that would lead to its occurrence. For instance, the customer usually has the best knowledge of the location of underground equipment at the site. Therefore, the customer should be responsible for identifying and marking the location of such equipment. Should the equipment be damaged, the customer is then responsible for absorbing the cost of the repair and any downtime costs.

Documentation. For a risk management process to be effective, it must be documented. The inclusion of limiting assumptions in a proposal can be an effective way of documenting and properly allocating risks. For example, if a customer has provided data that indicate that bedrock will be encountered at 15 feet, one of the assumptions in the proposal may state that the assumed depth to bedrock is 15 feet. If during the drilling operations it is discovered that the depth to bedrock is actually 35 feet, the project manager would be well within contractual rights to request a change order and not proceed unless one is granted.

It is important to realize that in estimating the scope, schedule, and cost of a job, one is constantly making assumptions. The importance of maintaining a running list of assumptions cannot be overemphasized. Including these assumptions in a proposal is an effective way of limiting risk.

Such written assumptions are also protective of the customer because they further define and place boundaries on the scope of work. The customer can better understand the exact costs that will be incurred. A vague response to an RFP, without limiting assumptions, can lead to misunderstandings down the road.

It should be understood that limiting assumptions are not the only type of response strategy to be employed. In fact, a number of strategies can be devised and documented, depending on the type of contract a customer expects the remediation contractor to sign. Types of contracts are discussed in the following section.

4.12.6   Contract Strategy Considerations

There are a range of possible types of contracts that can be employed on any project. The most common types used in the environmental remediation industry are

- Time and materials
- Firm fixed price
- Cost plus fixed fee
- Time and materials not to exceed

Allocation of risk is implicit in the type of contract selected.

Time and materials. A time and materials (T&M) contract allocates little or no risk to the seller. Under such a contract, the buyer agrees to pay whatever it takes to get the job done. If an event occurs that results in a significant cost increase, the customer must absorb the cost.

Firm fixed price. The firm fixed price (FFP) contract is the opposite of the T&M contract in terms of the allocation of risk. Here, the customer agrees to pay only the amount specified in the contractor's proposal — and no more. The contractor is assuming all the risk of something going wrong.

An FFP contract should be entered into only if the contractor has detailed knowledge of and experience with the job to be performed. Although this type of contract carries a high degree of risk for the contractor/seller, there is also the opportunity to realize windfall profits if the contractor can efficiently perform the job at a cost below that specified in the contract. The contractor still gets paid the contract amount even if the actual cost of the job is much less. Many construction contractors are skilled at developing proposals for FFP contracts that have a sufficient contingency budget which is taken as profit if very few risk events actually occur. In FFP contracts, it is important to develop detailed response strategies for the various risk events that may occur.

Cost plus fixed fee. The cost plus fixed fee (CPFF) contract provides reimbursement to the seller for all allowable costs plus an agreed-upon fixed fee which represents the seller's profit. This type of contract is used most frequently in contracts performed for the federal government.

Cost refers to the actual cost incurred by the seller in performing the project. Therefore, labor hours are charged not at a retail rate (which includes profit) but at the actual cost to the seller (including salary, benefits, etc.).

In order to accommodate a CPFF contract, the seller must be able to reconcile all costs incurred (actual, not retail) in performing each task (personnel, equipment, etc.). This can be quite difficult, particularly in terms of labor and corporate overhead. Depending on the terms and conditions of the contract, some of the incurred costs may or may not be chargeable to the customer (allowable).

Risk is shared on a relatively equal basis by the customer and the seller in a CPFF contract. The customer's risk is having to pay whatever it costs to complete the project. The seller's risk is a potentially reduced percentage profit. This results from the fact that profit is set at a fixed amount. Although the seller will be reimbursed for all costs (and therefore not lose money), the percentage profit earned is reduced as the costs increase. Risks inherent in the project can cause costs to increase to the point where the fixed fee (profit) is very small by comparison. In such a case, the seller may have lost the opportunity to assign personnel to projects that earn a greater profit.

Time and materials not to exceed contract. The time and materials not to exceed contract, which is most frequently used in the commercial environmental remediation industry, is theoretically the least favorable for the contractor. Under this type of contract, the customer will pay only for the cost of the work up to a limit established in the contract.

If the contractor can perform the work at a cost less than the preset amount, less money is earned. If the job costs more than the preset limit, the contractor must absorb the cost. Fortunately, most customers will allow for the limit to be exceeded if it can be demonstrated that the scope of work has changed.

Providing the customer with a proposal that contains a detailed scope of work and limiting assumption facilitates the process of demonstrating a change in scope (should one occur). This is because the detailed proposal explains exactly what items are and are not included in the project cost. The techniques taught in this book will assist in the development of detailed proposals.

Figure 4.4 presents a qualitative indication of risk allocation by contract type.

4.12.7 Overall Project Risks

All projects represent an opportunity for profit and contain inherent risk. This section has provided some guidance on how to identify risk within a project and manage it within the planning stage. Some projects by their very nature are so out of the ordinary that they are fraught with risk (like landing someone on the moon). Oftentimes, we are intrigued by a project and want to be the first to accomplish the challenge. There is nothing wrong with this desire, but some rules of thumb may be helpful in deciding whether or not to proceed. An opportunity may not be worth the risk if:[19]

- All risk associated with the project is allocated to the contractor through a mechanism such as an FFP contract.
- Technical personnel indicate there is less than a 50% chance of success.
- Due to financial pressures, the operations unit simply cannot afford to lose money.
- The benefits cannot be identified.
- There simply are not enough data to formulate a strategic plan (in which case more research is needed).
- There are other projects available which have a better chance for success and the potential for a greater profit margin.

4.13 THE FINAL BID–NO BID DECISION

In reviewing the RFP thus far, we have looked at:

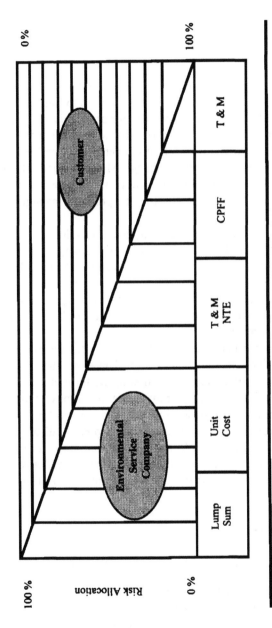

FIGURE 4.4 Risk allocation by contract type.

- Summarizing data
- Identifying objectives
- Identifying specifications
- Analyzing stakeholders
- Evaluating project risks

Now we are in a position to make the final bid–no bid decision. To make the final bid decision, the proposal manager is required to finish filling out Sections E to H of the Final Bid–No Bid Evaluation Form. This will require estimating the proposal preparation cost. The cost estimate will be based on estimating the hours that will be involved for each of the personnel involved in preparing the proposal. If possible, the proposal manager should contact all individuals who are expected to be used in the production of the proposal and ask them to estimate their hours to perform the tasks the proposal manager has planned for them.

Once this is done, the proposal manager can develop a preliminary estimate of the gross revenue and expected profit based on margin percentages provided by operations management. Estimated profit can then be calculated by multiplying the gross revenue by the margin (in decimal percent).

Finally, the estimated profit is multiplied by the probability of award and divided by the preparation cost. This gives the profit to proposal cost ratio, which should be greater than 1.0

It should be noted that Sections E to H of the Bid–No Bid Evaluation Form require an extensive amount of estimation. Therefore, the final bid–no bid decision should not be based on this portion of the form alone. The final bid– no bid decision should be a joint effort involving the proposal manager, operations management, contracts administration, and sales. A meeting or conference call should be organized, and the core team should present all of its findings to the other participants. Summary information and completed forms should be provided to the participants early enough to allow them time for review prior to the call. The final decision can then be made during this meeting.

4.14 EXERCISES

This chapter involves three exercises pertaining to the Megacorp site:

1. Fill out page 1 of the Bid–No Bid Evaluation Form.
2. Complete the RFP Summary Form.
3. Complete the Stakeholder Management Summary Form.

Try to fill out these forms as completely as possible. This experience will provide you with a great deal of insight into the project and the usefulness of the forms. By filling out the forms, you should have a better understanding of the project uncertainties which are creating risks. The following issues should become apparent as a result of completing the forms:

1. The extent of groundwater impact is not fully known.
2. The last groundwater sampling event was in 1990.
3. It is uncertain if the two sand lenses in the underground storage tank area are in communication with each other and with the sand lens in the former disposal area.
4. Megacorp's corporate environmental lawyer and environmental manager have been given conflicting objectives by senior management.
5. The two areas of site impact are distinct in the ways in which contaminants were released, types of contaminants, and ownership.

Many companies that have been exposed to these forms utilize them in their overall planning process and keep the completed forms in project files and computer records. The forms, along with historical project plans and results, can be retrieved for review when planning future projects.

Appendix F contains the completed forms for this exercise. You may choose to compare them with your own results. You may also identify additional issues not included by the author.

REFERENCES

1. Meador, R., *Guidelines for Preparing Proposals: A Manual on How to Organize Winning Proposals for Grants, Venture Capital, R&D Projects, Other Proposals*, Lewis Publishers, Boca Raton, FL, 1988, 4.
2. Larsen, H.T., On the value of cost estimating accuracy under conditions of competitive bidding, *Cost Engineering*, Vol. 36, 25, 1994.
3. American Association of Cost Engineers, Education Board, *Skills and Knowledge of Cost Engineering*, 3rd ed., AACE, Morgantown, WV, 1992.
4. Cleland, D.I., *Project Management, Strategic Design and Implementation*, McGraw-Hill, New York, 1990, 174.
5. Schuyler, J.R., Decision analysis in projects, modeling techniques. 1, *PM Network*, Vol. VII, No. 7, 26, 1994.
6. Ireland, L.R., *Quality Management for Programs and Projects*, Project Management Institute, Drexel Hill, PA, 1991, II-5.
7. Pinto, J.K., Successful project management: do you know your politics? *PM Network*, Vol. VIII, No. 7, 33, 1994.

8. Eccles, R. et al., *Beyond the Hype,* Harvard University Press, Cambridge, MA, 1992.
9. Cleland, D.I., *Project Management, Strategic Design and Implementation,* McGraw-Hill, New York, 1990, 95.
10. Cleland, D.I., *Project Management, Strategic Design and Implementation,* McGraw-Hill, New York, 1990, 103.
11. Cleland, D.I., *Project Management, Strategic Design and Implementation,* McGraw-Hill, New York, 1990, 106.
12. Cleland, D.I., *Project Management, Strategic Design and Implementation,* McGraw-Hill, New York, 1990, 107.
13. Cleland, D.I., *Project Management, Strategic Design and Implementation,* McGraw-Hill, New York, 1990, 111.
14. Cleland, D.I., *Project Management, Strategic Design and Implementation,* McGraw-Hill, New York, 1990, 109.
15. Wideman, R.M., *Project and Program Risk Management: A Guide to Managing Project Risks and Opportunities,* Project Management Institute, Drexel Hill, PA, II-3.
16. Project Management Institute Standards Committee, *Project Management Body of Knowledge (PMBOK),* Project Management Institute, Drexel Hill, PA, 1987, E-2.
17. Fraser, D.C., Risk minimization in giant projects, paper presented at International Conference on the Successful Accomplishment of Giant Projects, London, May 1978.
18. Wideman, R.M., *Project and Program Risk Management: A Guide to Managing Project Risks and Opportunities,* Project Management Institute, Drexel Hill, PA, II-3.
19. Wideman, R.M., *Project and Program Risk Management: A Guide to Managing Project Risks and Opportunities,* Project Management Institute, Drexel Hill, PA, I-7.

Assembling Project Teams

5

Once a proposal has been thoroughly reviewed and a decision has been made to bid on a project, detailed planning can begin, with the exception of one factor: assembling the project team.

The issues that must be considered in detailed planning of an environmental remediation project are multidisciplinary in nature; therefore, a project should be planned by a multidisciplinary team. Identifying key team members early on will allow their constructive input into planning phases such as development of the work breakdown structure, network diagram, and cost estimate.

Project management is designed to put the best mix of people together in order to achieve project objectives. More than all other factors, the success of a project will depend directly upon:

- The assignment of personnel to tasks which are commensurate with their skills
- The willingness of individuals to diligently apply their skills to the completion of assigned tasks

Since team performance is crucial to project success, this chapter focuses on:

- Identifying key project personnel
- Gaining commitment
- Performing team planning
- Developing the project organizational chart
- Determining the core project team and its location

It should be noted that this chapter deals with identifying and gaining the commitment of **key project personnel**, those individuals who will be responsible for either performing or managing major project tasks. The resumes of these key project personnel will be included in the project proposal. Other project team members whose functions are less significant are discussed in Chapter 8.

5.1 IDENTIFYING KEY PROJECT PERSONNEL

The identification of key project personnel requires an understanding of:

- The major tasks which are likely to be performed, in order to reach the project objects
- The types of skill sets (disciplines) required to perform these tasks
- The availability of personnel who have these required skill sets

5.1.1 Major Tasks to Achieve Project Objectives

Performing a detailed preliminary review of a project as discussed in Chapter 2 should provide a great deal of insight into the primary tasks required. In most cases, these tasks will be identified by the project objectives (including long-term, short-term, and subobjectives). Additionally, the major tasks associated with the project stages are often identified in guidance documents for whichever regulatory program is driving the project (e.g., RCRA, CERCLA, etc.). For example, "Guidance for Conducting Remedial Investigations and Feasibility Studies under CERCLA" (OSWER Directive 9355.3-01) (RI/FS guidance document) indicates that there are four major tasks in performing a feasibility study.[1] This RI/FS guidance document also provides information on the subtasks associated with each major task.

The major tasks and subtasks as provided by the RI/FS guidance document are

1.0 Development of Alternatives
    1.1 Identify Potential Treatment/Containment Technologies
    1.2 Screen Technologies
    1.3 Identify Action Specific Applicable, Relevant, and Appropriate Requirements
    1.4 Assemble Technologies into Alternatives
2.0 Screening of Alternatives
    2.1 Further Define Alternatives
    2.2 Evaluate Defined Alternatives (Effectiveness, Implementability, Cost)
    2.3 Screen Alternatives
    2.4 Verify Action Specific Applicable or Relevant and Appropriate Regulations

3.0 Treatability Testing
    3.1 Bench Scale Testing
    3.2 Pilot Scale Testing
4.0 Detailed Analysis of Alternatives
    4.1 Further Define Alternatives (As Necessary)
    4.2 Analyze Alternatives Using Nine Criteria
    4.3 Compare Alternatives Against Each Other

Although a project, as defined in a request for proposal (RFP), may encompass completion of only one or two project stages, including on the planning team personnel who are associated with later stages will facilitate comprehensive planning. For instance, design personnel could be included in the planning of a feasibility study, since they might recommend gathering data that will facilitate future design efforts.

5.1.2 Skill Sets Required to Perform Tasks

Once the major tasks are identified, the skills sets required are usually apparent. If a project calls for a groundwater fate and transport model, it is obvious that a hydrogeologist experienced in modeling will be required. In a similar fashion, if a significant amount of regulatory interpretation and negotiation is required, a skilled regulatory specialist should be included on the project team.

5.1.3 Typical Project Team

To assist further in team identification, the typical project team for a large-scale environmental remediation project would consist of the following positions:

- Project director
- Project manager
- Regulatory specialist
- Remediation engineer
- Project geologist/hydrogeologist
- Risk assessment manager
- Feasibility study engineer
- Construction engineer
- Design engineer
- Project management assistant

The last position listed above, the project management assistant (PMA), refers to an individual who has an administrative background. The PMA assists in developing project status reports (budgets, schedule), maintaining project files, and documenting project communications.

Note that the role of the PMA should not be confused with that of a task manager, functioning in a transitional role from technical specialist to project manager. Individuals in such a transitional role would work closely with the project manager in planning, monitoring, and controlling project execution (rather than functioning as a PMA).

5.1.4 Availability of Personnel

One of the most difficult problems faced in assembling a project team is finding skilled individuals who are available to fill the various project positions. Typically, the more skilled an individual is, the greater the demand for his or her services. It can safely be said that all project managers (and customers as well) want only the most skilled individuals working on their projects. However, this is not always practical.

For marketing purposes, it is common in nearly all consulting firms to include the resumes of highly skilled individuals in a project proposal. There is nothing wrong with this practice since it honestly presents the capabilities of the company. However, care should be taken in presenting the role and responsibilities of these highly skilled individuals.

For example, suppose a remediation service company has just received an RFP for a small project on the East Coast involving *in situ* bioremediation, but the most experienced *in situ* bioremediation expert works out of an office located on the West Coast. From a marketing standpoint, it would make sense to include the bioremediation expert's resume in the proposal. However, due to availability and cost, it may not be practical to make this person responsible for managing the project or performing any major tasks. Suggesting to the customer that the bioremediation expert would play a major role in project execution may be misleading and could cause credibility problems at a later point in time. One way to take advantage of an expert's experience is to present him or her as a technical consultant. The text of the proposal could explain that this person has provided input in planning the project scope of work and that he or she will be available for technical consultation during execution.

5.1.4.1 Geographically Distributed Project Teams

Advances in communication technologies and transportation services have made it possible to move information and materials in time frames that could not even be imagined 20 years ago. As a result of these advances, it has become increasingly possible for individuals to provide services from remote locations. Over the past decade, throughout most industries, it has become common for project teams to be geographically distributed. Remediation service companies are no exception to this trend.

Although it is much more challenging to plan and manage a project using geographically distributed team members, one primary advantage is that it allows project managers to utilize skilled personnel who might not otherwise be considered. This permits a company with a number of offices located throughout the United States and abroad to perform large complex projects which at one time would have been beyond the capabilities of a local office.

The difficulties of geographically distributed project teams include:

- Being aware of all the personnel who could potentially fulfill the project requirements
- Evaluating the skill level of personnel offered by various operations managers as potential team members
- Gaining and maintaining the commitment of project personnel

Methods of overcoming these disadvantages are discussed in the next several pages.

5.1.4.2   Recognizing Available Personnel

In a large company that consists of many offices, it is difficult to know all the individuals who could assist on a project. With the advent of wide area networks, e-mail, centralized databases, etc., project managers can retrieve the resumes of individuals who have particular skill sets. Even though high-tech methods of gaining access to company resources may be available, project managers should not underestimate the value of direct contacts. Operations and marketing managers are in positions that expose them to a wide variety of company personnel. The same is true of project directors. If the project review process discussed in Chapter 2 has been followed, one or more of these managers/directors should be well aware of the project and able to offer suggestions for personnel to fill the team positions.

Ultimately, the availability of a certain individual should be determined through discussions with that person's direct supervisor (operations or line manager). Usually, the supervisor will want an estimate of the hours per week that the individual may be required to devote, the number of weeks, and the actual time period. Such an estimate may be difficult to provide at the early stages of planning. However, once the network schedule has been designed and resources have been assigned, this question will be much easier to answer. Because these personnel will be used to assist in designing the network schedule and estimate resource requirements, the project manager may have to provide a rough estimate to be revised later.

If the project manager would like an additional measure of security regarding the availability of a certain individual, he or she may want to speak directly to that individual. However, if the operations manager is performing his or her job effectively, this direct communication may not be necessary.

5.1.4.3 Evaluating Skill Levels

Once a list of potential project team members has been developed, the project manager must be sure that these individuals do indeed possess skills at the level demanded by the project. Note that the skill level demanded by a project may be less than what the project manager would like a potential team member to possess. The project manager must remember that it is not always necessary to assemble the "A Team" of environmental remediation.

In some cases, personnel may be offered by operations managers based on availability. In a consulting firm, availability often equates with low billability. The project manager should keep in mind that although operations managers may be attempting to provide genuine assistance, they might also be concerned with maintaining the billability of the individuals who report to them. There are a number of reasons why an individual may have low billability, including time spent on proposals, low volume of work at the local office, and lack of name recognition throughout the company. However, in some cases, an individual can have low billability due to a demonstrated lack of skills. If an individual's skill level is not up to par, the operations manager must see that appropriate training is provided. The best way to investigate the skill level of a potential team member, other than discussions with his or her operations manager, is to contact other project managers who have had previous experience with this person.

5.2 GAINING AND MAINTAINING COMMITMENT

According to much of the project management literature, the number one cause of project failure is lack of committed resources. Lack of committed resources refers, in general, to the personal commitment of the team member. However, in a some cases, it can also refer to a simple lack of necessary resources. In any discussion among project managers, regardless of industry background, this lack of committed resources is a common theme.

An informal survey of project managers was performed during a project management training session conducted by a major environmental service company in May 1990. Each project manager was asked to provide a list of frequently encountered problems. The survey indicated that, consistent with project management literature, lack of committed resources was the most frequently encountered problem.

5.2.1 Commitment and Leadership

The question of gaining and maintaining commitment relates to the concept of leadership. Although the subject of leadership has received much attention, a universally accepted definition of the term has not been identified. In the book *Leadership*, by James McGregor Burns, one study with over 130 definitions of the term is cited.[2]

Thousands of articles which explore leadership traits have been published. Some of the traits are said to relate to abilities, some to physical characteristics, many to personality, and others to social characteristics.[3] With so much investigation yet so little agreement, a comprehensive study of leadership will not be undertaken here. However, a definition of leadership in the project setting has been selected, and practical suggestions regarding leadership style are provided.

5.2.1.1 Project Definition of Leadership

For the purposes of project management, our accepted definition of project leadership is as follows:

> **Project leadership** is the ability to persuade others to enthusiastically and diligently pursue the project's objectives.[4]

5.2.1.2 Leadership Style

The preceding definition of project leadership leads to a question: *How does one develop the ability to persuade others to pursue the project objectives?* It is at this point that much of the project management literature leaps into a discussion of leadership style. The numerous articles and books on the subject, which often present contradicting viewpoints, can give the impression that no one knows what an effective leadership style looks like.

Much of the literature on leadership looks at situations where the intended objectives were met and then discusses the style that was employed in these cases. The problem, however, is that different styles work in different situations. Therefore, the logical conclusion is that *leadership is situational.*

Effective leadership depends on who is being led, who is doing the leading, and the environment in which they are operating. A highly useful article that discusses situational leadership, entitled "Project Leadership: Understanding and Consciously Choosing Your Style," by Dennis P. Slevin and Jeffrey K. Pinto, was presented in the March 1991 issue of the *Project Management Journal.* This article is provided in Appendix D for those who wish to delve into this subject further.

Briefly, the article discusses four extremes in leadership style, represented as the four corners of a grid. The authors further explain that almost every type of leadership style can be plotted in such a system. According to Slevin and Pinto, the four extremes in leadership style are as follows:[5]

- **Autocrat.** These types of leaders typically solicit little or no information from their project team and keep decision-making authority to themselves.
- **Consultive autocrat.** This leadership style features managers who seek intensive information input from the project team members, but who ultimately keep substantial decision-making authority to themselves.
- **Consensus manager.** This type of leader opens the problem up to the group for discussion (information input) and simultaneously encourages the entire group to make the management decision.
- **Shareholder manager.** This type of leader is the only one on the grid who is considered to represent poor management. Here, little or no information is exchanged within the group context, yet the group is provided with the ultimate authority for decision making.

Under stressful conditions, where time is of the essence, the autocratic style may be most appropriate. On the other hand, the consensus management style may be appropriate when decision making is controversial or will affect the group as a whole and/or when productivity is of less concern. Obviously, from the above discussion, shareholder management may never be appropriate. This leaves consultive autocratic management.

Consultive autocratic management is usually ideal for most project management settings. This type of management is appropriate when a project involves complex issues — yet time is of the essence. With consultive autocratic management, project team members are given the opportunity to voice concerns, present issues, and offer solutions. However, once the information has been gathered, the project manager steps in to decide the most pragmatic course of action, thereby maintaining productivity.

5.2.2 Team Commitment and Motivating Factors

The discussion of leadership styles indicates that in order to be successful (which entails having committed resources), leaders must be flexible and remain aware of the needs and motivations of team members, given the circumstances at hand.

A project manager within a remediation service company is, for the most part, managing a group of technical professionals. Although individuals are motivated by a wide range of needs and desires, it should be noted that technical professionals appear to have some common denominators in terms of motivation. The project management literature cites the following motivators for technical professionals:

- Involvement in the decision-making process (i.e., the ability to influence one's destiny)
- Interesting and stimulating work
- Recognition for accomplishment

Involvement in the decision-making process. Managing technical professionals involves dealing with a group of well-educated people who know how to logically analyze and solve problems. It is only natural for these individuals to have a desire to participate in the decision-making process. The project management literature is riddled with examples which indicate that professionals are much more likely to commit to a project when they have had an opportunity to participate in the planning decisions. It is clear that if you want a professional's commitment, you must involve him or her in the planning.

It should be noted that this does not mean that a project manager needs to act on every suggestion offered by technical professionals. On the contrary, it appears that professionals will commit even if their suggestions are not acted upon — as long as they are provided the opportunity to offer input. This is especially true if the project manager takes the time to provide a logical explanation whenever a suggestion is not acted upon, which is consistent with the style of consultive autocratic management.

Desire for interesting and stimulating work. Due to the technical complexity associated with most environmental remediation projects, finding interesting and stimulative work is usually not a problem. However, to enhance commitment to a project, a project manager may want to point out aspects of the project that make it particularly challenging or unique to prospective team members. In some cases, a project manager may show that some of the aspects of the project can provide a team member with experience that can enhance and round out his or her career background.

Desire for recognition of accomplishment. Recognition for accomplishment can take many forms. In some cases, it is as simple as thanking a team member for a job well done. Other forms of recognition include submitting an article to the company newsletter and congratulating an individual through a case study presentation at a company function.

Another type of recognition is a promotion. Unfortunately, however, a project manager is usually not in the position to directly promote a team member. Nevertheless, a project manager can certainly influence such a promotion by communicating accomplishments to the proper operations manager. Indeed, all project managers should make it a policy to inform operations managers of team members' performance — and make a special note when an employee has performed particularly well.

This discussion has focused on positive incentives to gain and maintain commitment, which, in most cases, is preferable. However, there are times when negative incentives are appropriate. If a team member is not performing in accordance with requirements and capabilities, he or she should be informed promptly by the project manager and given an opportunity to make adjustments. If adjustments are not forthcoming, an appropriate step would be to inform the operations manager of the difficulties. Furthermore, due to time constraints, the project manager may need to replace the team member with another qualified person, rather than wait for the original person to be properly trained.

5.3 TEAM PLANNING AND GEOGRAPHICAL DISTRIBUTION

Many of the suggestions for gaining commitment may seem to depend on the team members working in one location. Clearly, if all team members were assembled together in one conference room, collaborative planning would be streamlined. The International Project Management Association promotes such a personal approach with a technique called the Project Start-up Process. According to this process, the key members of a project team come together and in a three-day period jointly develop the project work breakdown structure, network diagram, and linear responsibility chart.

However, it would be a luxury in most remediation service companies for personnel to have the opportunity to plan a project in this ideal fashion. If the potential project is large (greater than $500,000 gross revenue) and the chance of winning is high (greater than 70%), planning a project in this fashion may be appropriate. For other projects, however, creative methods of team planning need to be identified.

One way is to have each of the key members develop a work breakdown structure, network diagram, and project budget for their specific portion of the project (task). Each task planned in this fashion would represent a subproject, similar to a subroutine in a large computer program. After reviewing and discussing these subprojects with each of key team members, the project manager could link the subprojects together within the project management information system, thereby compiling one full project. With the advent of e-mail, each of the key team members could plan their portion of a project at a remote location and then e-mail their planning file to the project manager.

Another way to approach geographically distributed teams is to have the project manager plan out each of the tasks and then provide copies to the respective task managers for review and discussion. Each task manager could then offer suggestions to make the plan more accurate.

Other creative methods are likely to be developed by individual companies. Through use of the Internet, wide area networks, e-mail, fax machines, voice mail, express delivery services, etc., there are a large number of ways to accomplish team project planning.

5.4 PROJECT ORGANIZATION CHART

Once we have identified the team positions, found personnel to fill the positions, and gained their commitment, we are ready to develop the project organization chart. The purpose of an organization chart is to indicate the authority and reporting structure of a company or team.[6] The traditional organization chart presents vertical relationships and has a pyramidal layout. These pyramidal charts are often criticized by sophisticated observers because they do not represent the horizontal and diagonal relationships that exist within any organization. Basically, the traditional pyramidal organization chart fails to deal with the nuances of informal control and authority and does not address those relationships which seem to develop naturally and which may, in fact, be necessary to the smooth operation of any group.

When developing the organization chart for a project, we are faced with two problems that make it difficult to clearly present authority and reporting structure. The first problem is that the project team is a subset of the total organization. This means that, in addition to reporting to the project manager, team members are required to report vertically to other company personnel (operations managers) who are not intimately involved with the project. In addition, from a total organization viewpoint, some project team members may have positions with much higher authority and responsibility than the project manager. The second problem in developing the organization chart is that the project management approach is based on the concept of replacing traditional vertical (and often bureaucratic) relationships with a flexible project team that can rapidly adjust to changing project conditions. This suggests a more fluid chart than is typically used in traditional management.

Historically, within both the project management literature and remediation service companies, attempts have been made to develop a project organization chart that acknowledges the nuances in communication and authority that exist within a project, while respecting a company's organizational structure. Early attempts involved the use of a spider diagram (Figure 5.1).

As indicated in Figure 5.1, the project manager is placed in the center and is encircled by the various members of the project team. The advantage of this type of project organization chart is that it reflects the fact that the project manager acts as the integrator of the project team, with all lines of communication open to the project manager. Unfortunately, however, this appears to be the only advantage to this type of organization chart.

Experience within various remediation service companies indicates that the two biggest problems with spider diagrams are that (1) customers find them confusing and (2) authority relationships are not clearly identified. Some of the confusion may have resulted from the fact that customers simply were not used to seeing an organization chart presented in this manner. A further explanation for the customer confusion is that the spider diagram indicates only horizontal lines of communication — and not authority relationships. For example, does the field geologist report directly to the project manager or to the project hydrogeologist? Does a central engineering group review the work performed

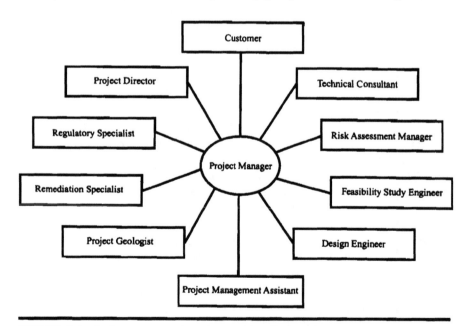

FIGURE 5.1 Project organization chart: spider diagram format.

by the feasibility study engineer, or is the work reviewed entirely by the project manager?

To correct the problems associated with spider diagrams, many project managers develop combination charts (i.e., mix a pyramidal chart with a spider diagram). In these charts, the project manager is still placed in the center. However, in the combination chart, a box might be placed below the project hydrogeologist to indicate authority over the field geologist. To indicate a review function, a dotted line might be drawn above the feasibility engineer leading to a central engineering group. In some cases, however, these combination charts are inaccurate. One common error, for instance, is placing the operations manager between the customer contact and the project manager. This suggests that customer communications should flow through the operations manager and not the project manager. Another problem with combination charts is that they sometimes are so complex as to be comical.

5.4.1 Project Organization Chart and the Customer

A point to remember about project organization charts is that, to a great extent, they are prepared for the benefit of the customer. With this in mind, the project organization chart should be as simple as possible. For the sake of simplicity, the traditional pyramidal chart still seems to be the most practical; therefore, it is recommended for most remediation service companies. An example of a project organization chart is presented as Figure 5.2.

5.4.2 The Linear Responsibility Chart

The preceding discussion should make it clear that no matter how they are designed, most organization charts seem to lack some information regarding lines of communication, authority, and responsibility. It is for this reason that linear responsibility charts were developed. David Cleland and William King appear to be among the first to promote the use of linear responsibility charts.[7] Linear responsibility charts capture the information in traditional organization charts and associated personnel manuals and combine the data in a matrix format.

The linear responsibility chart (matrix) is developed by combining the project organization chart with the work breakdown structure. The concept of the work breakdown structure is discussed in detail in Chapter 6. The linear responsibility chart is discussed briefly in this section and will be revisited later in Chapter 8.

Briefly, the linear responsibility chart provides a series of position titles along the top of a table (columns). A list of tasks, activities, functions, etc. is then presented down the left side as row headings. Finally, an array of symbols

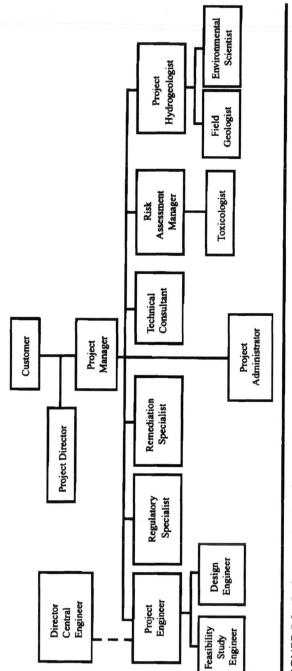

FIGURE 5.2 Project organization chart: traditional format.

indicates the extent of authority and the relationship between the columns and lines. A small example of a linear responsibility chart is provided in Figure 5.3.

The linear responsibility chart attempts to answer such questions as: Who has signature authority? Who must be notified? Who can make the decision? These questions can only be answered by clear definitions of authority, responsibility, and accountability.

Definitions offered by Kerzner are provided below:[8]

- *Authority* is the right of an individual to make the necessary decisions required to achieve his or her objectives and responsibilities.
- *Responsibility* is the assignment for the completion of a specific event or task.
- *Accountability* is the acceptance for success or failure.

Some organizations report that linear responsibility charts are highly valuable, since they show, at a glance, who is responsible for which functions and how certain positions relate to other positions. At a workshop given during the 1992 Seminar/Symposium of the Project Management Institute, a group of project managers from AT&T presented several case studies on how capital construction projects were enhanced by the use of linear responsibility charts. Their use is recommended for remediation service companies as well.

5.5 THE PROJECT OFFICE AND LOCATION

In companies with large capital construction projects, an organization known as the project office is often developed. The project office is designed to support project managers in carrying out their duties. Project office personnel must have the same dedication to a project as the project manager — and have a good working relationship with the various task managers.[9] According to Kerzner, the responsibilities of the project office include:[10]

- Acting as the focal point for information for both in-house control and customer reporting
- Controlling time, cost, and performance to adhere to contractual requirements
- Ensuring that all required work is documented and copies are distributed to all key personnel
- Ensuring that all work is both authorized and funded by contractual documentation

Operations and Team Resources

Activities / Function	Project Director	Project Manager / Task Manager	Project Management Assistant	Risk Assessment Manager	Director Risk Assessment	Toxicologist	Project Hydrogeologist	Field Geologist	Environmental Scientist	Project Engineer	Feasibility Study Engineer	Design Engineer	Director Central Engineer
Project Objectives	●	△		△		▲	△	▲		△	△	△	▲
Monthly Status Reports	■	●	●	▲		▲	▲	▲		▲			
Fate & Transport Model	✱	●			△	▲	●						
Screening of Technologies	■	✱					△			■	●	▲	■
Document Turn-around	■	●	○	○		○	○	○	○	○	○	○	
File Maintenance	▲	△	●							■	△		
Pilot Test Design	■	▲				▲	△	▲		■	△	●	■

Legend

Symbol	Meaning	Symbol	Meaning
■	Must Approve	△	Will Be Consulted
●	Specialized Responsibility	□	Must Issue
▲	May Be Consulted	✱	Must Review
		○	Will Receive

FIGURE 5.3 Linear responsibility chart.

5.5.1 The Core Project Team

It is obvious that such a project office would be highly beneficial to any project manager. However, in the past, an extremely large project was required in order to justify such a large organization. Given the organizational structure of many remediation service companies (geographic distribution) and the size of most environmental remediation projects (even the larger ones), such a project office might seem out of the question. This does not have to be the case.

By taking a larger view, it can be seen that the project manager, transitional task manager(s), and PMA together can carry out the functions typically associated with the project office. These individuals represent the core project team. Together this core team can be responsible for the management of one large project or many small projects.

5.5.2 Core Team Location

Development of a core team is recommended as a way of coping with geographically distributed personnel. To ensure that the individuals on the core project team have the same dedication as the project manager, all members of the core team should be located in the same office. It is also recommended that the core project team be located in the office that is geographically closest to the customer contact. Close proximity will facilitate project status meetings with the customer, thereby improving overall communication and visibility.

The assignment of a core project team to all projects will ensure that two other individuals (transitional task manager and PMA), in addition to the project manager, are intimately familiar with the project in terms of status reporting and data management. A further advantage of a core project team is that new project managers can be developed through close contact with an experienced project manager.

5.6 EXERCISES

This chapter involves two exercises pertaining to the Megacorp site:

1. Develop a project organization chart that presents key project team personnel.
2. Construct a linear responsibility chart for the bid responses based on the information contained in the Megacorp Site Summary (Appendix A).

Recall that the project organization chart is prepared for the benefit of the customer and the linear responsibility chart is prepared for the benefit of the internal project team and operations management. Developing the project organization chart will require identification of the major tasks to achieve project objectives. Specific project objectives will be developed as part of the exercises in Chapter 6. For now, however, those attempting to develop a project organization chart may find it helpful to review Section 13.1 of the Megacorp Site Data Summary (Appendix A). This section indicates major tasks such as investigation, pilot testing, and design. In addition, one could surmise that determining which technology to pilot test could involve a feasibility study and perhaps even a risk assessment to develop cleanup goals.

The row headings should contain tasks such as complete RFP Summary Forms and Stakeholder Management Summary Form.

The reason why the linear responsibility chart exercise involves proposal responsibilities, rather than project responsibilities, is because the exercise provides familiarity with the linear responsibility chart technique, while at the same time requiring individuals to think through the responsibilities key team members should perform in executing the planning activities of the project planning and control cycle (see Figure 3.13). Recall that proposals are projects. Many companies may wish to develop a standardized linear responsibility chart to guide proposal preparation. Such a chart should improve communication during proposal preparation, help reduce proposal response time, and eliminate duplication of effort.

Appendix F contains a sample project organization chart and linear responsibility chart that could be created in completing this exercise.

REFERENCES

1. Office of Solid Waste and Emergency Response, U.S. Environmental Protection Agency, Directive 9355.3-01, Draft Guidance for Conducting Remedial Investigations and Feasibility Studies Under CERCLA, U.S. Environmental Protection Agency, Washington, D.C., 1988, 4-1, 5-1, 6-1, 7-1.
2. Burns,, J.M., *Leadership*, Harper & Row, New York, 1978, 2.
3. Cleland, D.I., *Project Management, Strategic Design and Implementation*, McGraw-Hill, New York, 1990, 255.
4. Davis, K., *Human Relations at Work*, 3rd ed., McGraw-Hill, New York, 1967, 96.
5. Slevin, D.P. and Pinto, J.K., Project leadership; understanding and consciously choosing your style, *Project Management Journal*, Vol. XXII, 41, 1991.
6. Cleland, D.I. and King, W.R., Linear responsibility charts in project management, in *Project Management Handbook*, 2nd ed., Cleland, D.I. and King, W.R., Eds., Van Nostrand Reinhold, New York, 1988, 375.
7. Cleland, D.I. and King, R.W., *Systems Analysis and Project Management*, 3rd ed., McGraw-Hill, New York, 1983.

8. Kerzner, H., *Project Management: A Systems Approach to Planning, Scheduling, and Controlling*, 2nd ed., Van Nostrand Reinhold, New York, 1984, 235.
9. Kerzner, H., *Project Management: A Systems Approach to Planning, Scheduling, and Controlling*, 2nd ed., Van Nostrand Reinhold, New York, 1984, 192.
10. Kerzner, H., *Project Management: A Systems Approach to Planning, Scheduling, and Controlling*, 2nd ed., Van Nostrand Reinhold, New York, 1984, 193.

Developing a Work Breakdown Structure

6

The term work breakdown structure (WBS) has been mentioned a number of times in the previous chapters. In this chapter, the functions performed by the WBS, as well as methods for its development, are discussed in detail. The WBS is seen by many individuals who are new to project management as simply a tool for scope definition. However, experienced project managers understand that the WBS does much more than outline the scope of work; the WBS provides the means for project integration, communication, and control.

The primary topics discussed in this chapter include:

- A definition of the WBS
- Functions of the WBS
- Specifications for the WBS
- Techniques for developing the WBS
- Rules of thumb for evaluating the WBS
- Templates for constructing the WBS
- Scope of work and specification development based on the WBS
- Risk management based on the scope of work, specifications, and WBS

Note that WBS templates for generic environmental remediation projects are provided in Appendix E. These templates, designed specifically for the environmental remediation industry, serve as a starting point for developing a project-specific WBS.

6.1 DEFINITION OF WORK BREAKDOWN STRUCTURE

Because of the many functions performed by the WBS, it is difficult to define in a single sentence. The 1987 edition of the Project Management Institute's

Project Management Body of Knowledge (PMBOK) contained a definition of the WBS that was somewhat awkward and difficult to comprehend. A much better definition, although general in nature, was provided in the 1996 edition of the PMBOK. Anyone planning to take the Project Management Professional exam will need to know this definition.

However, for the purposes of environmental remediation, the following definition is proposed:

> A WBS is a task-oriented, project-specific "family tree" that organizes the total work required to achieve the project objectives — as broken down into manageable work packages — while graphically providing a common framework for project planning, monitoring, and communication.

6.2   WORK STRUCTURE BREAKDOWN FORMATS

WBSs can be presented in either tree diagram format or indented outline format. Most project management software programs permit developing and printing the WBS in either format. Both formats present the project hierarchy broken down into stages, tasks, subtasks, and finally work packages.

6.2.1   Tree Diagram Format

The tree diagram format is presented in Figure 6.1. The main benefit of the tree diagram format is that it graphically displays the project hierarchy. It is very easy to see how the work packages add up to form subtasks, subtasks form tasks, and so on, thereby describing the entire project.

The problem with a WBS presented in tree diagram format is that it can become too large to be useful. Presenting an entire RCRA or CERCLA project in tree diagram format could require many sheets of D-size drawing paper. Because of this, it is not common to see WBS printouts in tree diagram format in actual project management practice.

In looking at the tree diagram presented in Figure 6.1, note that life cycle level of effort curves have been superimposed on the WBS elements. This calls attention to the fact that lower level WBS elements (tasks, subtasks, subsubtasks, etc.) are projects in their own right (i.e., each has its own objectives and life cycle). Therefore, the WBS represents not only a hierarchy of tasks, but objectives as well.

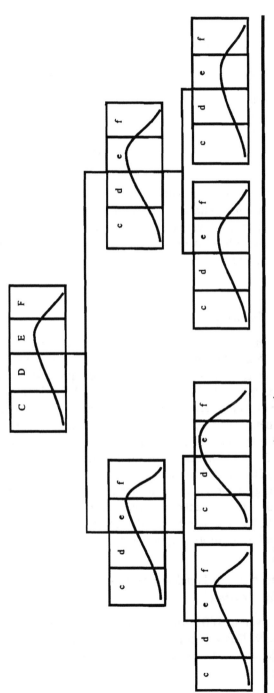

FIGURE 6.1 Work breakdown structure: tree diagram format.

6.2.2 Indented Outline Format

The following is an example of the indented outline format:

Level	Description
1	Project
2	Stage
3	Task
4	Subtask
5	Work package

A WBS with five levels is usually sufficient to outline most projects. Occasionally, in cases where a number of projects are being managed for the same customer, a new summary level is added, which is referred to as the program level. When a program level is required, level 1 would represent the program, level 2 the projects, level 3 the stages, and on down the line, resulting in a six-level WBS.

6.2.3 Cycling Format

A cycling WBS format is utilized on research projects that are performed in phases. Although WBS elements are usually thought of as representing unique, one-time events (in keeping with the definition of a project), research projects present a different approach. In this case, a cycling WBS is recommended because as tasks are executed, new information may create the need to repeat tasks at a new level of detail. This need to repeat a task more thoroughly is due to the low quality of information that existed when the task was originally planned. (See Section 3.8.2.7 for a discussion of the low quality of information during project planning.) Because of the low quality of information that is available during the planning of most environmental investigations, investigations are usually performed in phases, although every effort should be made to prevent the repetition of tasks. It is safe to say that most customers of environmental services would prefer not to look upon their sites as research projects.

6.2.3.1 Example of the Work Breakdown Structure Cycling Format

An example of a cycling WBS for an environmental remediation project was present in Figure 2.9. This was originally included in a quality assurance project plan (QAPP) for a large-scale RCRA Facility Investigation (RFI). It was used to present the general RFI tasks (physical characterization, interim remedial action, and analytical characterization) while also showing the flow of work through later project stages.

Figure 2.9 communicated to the Environmental Protection Agency that the master tasks of the RFI would be physical characterization, analytical characterization, and interim remedial actions. The physical characterization of the site would provide information on how the chemicals could be transported (dispersed) through the subsurface environment. The analytical characterization would provide information on the nature and extent of contamination. The interim actions (which for this RFI include a series of technology pilot tests) would provide additional physical and analytical characterizations.

Once the RFI tasks were completed for a given phase, Figure 2.9 indicated the order of the remaining project stages. In this case, a risk assessment would be performed, followed by a corrective measures study. Upon completion of the corrective measures study, one of the following choices would be selected:

- No further action
- Corrective measures design, followed by implementation, operation, and eventual closure
- Additional information required

If it was determined that additional information was required, a new cycle through the WBS would be required (a new phase). New tasks would be performed to further characterize the site physically or analytically. The risk assessment and corrective measures study would be revised as necessary. The cycling process would be repeated until EPA approval was achieved for an appropriate remediation technology or for no further action.

6.3 WORK BREAKDOWN STRUCTURE TERMINOLOGY

WBS terminology can be confusing. The confusion stems from:

- The fact that a number of terms are used interchangeably
- Past disagreement within the project management profession regarding WBS naming conventions

6.3.1 Interchangeable Terms

To understand the problem of interchangeable terms, it is helpful to refer to the WBS tree diagram format presented in Figure 6.1. Note that a number of terms are used interchangeably to describe both the boxes that make up the WBS as well as the relationships among the boxes. For instance, the terms task, project, and element are all used interchangeably to describe the boxes that make up the WBS. Depending on relationships within the WBS, the boxes are additionally referred to as:

- Stages, tasks, subtasks, and work packages
- Projects, subprojects, and subsubprojects
- Upper level elements and lower level elements
- Summary elements and constituent elements
- Parent tasks and child tasks

Note that one set of terms is not recommended over another. What is more important is that the relationships of the boxes are understood and that work represented by the boxes adds up to create the total project. Some terms seem to sound better, depending on the context within which the WBS is being discussed. Therefore, the terms will be used interchangeably in this chapter. When talking in general about the boxes that make up the WBS, the term elements will be used most often.

6.3.2 Naming Conventions (The Deliverables Issue)

There has been some disagreement among project management professionals regarding the actual naming of the WBS elements. The argument seems to be centered around the belief among some project managers that the WBS should only call out deliverables.

A **deliverable** is defined as any measurable, tangible, verifiable item that must be produced to complete a project.[1] This definition is often interpreted as meaning **objects** (i.e., things that can be picked up, manufactured, delivered, pointed to, or assembled). A WBS created by project managers who subscribe to the "deliverables only" approach contains only nouns for each element, such as *building* or *foundation*, and never uses words that indicate action (verbs), such as *install, collect,* or *drill.*

Project managers who believe that the WBS should only call out deliverables would say that a task which must be performed but which has no defined deliverable, such as administration, should not be included in the WBS. The problem with this thinking is that some basic tasks would be left out of the WBS — in particular, project management itself!

It is important, when choosing WBS nomenclature, that the names be easily understood by all project participants. The deliverables approach to naming WBS elements is neither necessary nor a hard and fast rule. (Some individuals, however, may still find it helpful to think their way through the project in terms of deliverables.)

6.4 WORK BREAKDOWN STRUCTURE DETAIL

Increasing the number of levels within the WBS will expand the level of detail at which the project is planned and controlled. However, all experienced project managers know that there is a point of diminishing returns in breaking down a project. Conversely, neglecting to break the project down far enough will make estimating and accurate tracking very difficult. Finding the proper level of detail requires careful thought and experience, although the guidelines provided later in this chapter can assist in this matter.

Generally, projects are usually planned and managed with a high degree of detail, while reports to upper level management and the customer are prepared at a lower level of detail. For example, Figure 6.2 presents the expanded WBS for a physical characterization task (based on the WBS presented in Figure 2.9). This expansion indicates the degree of detail to which the physical characterization would be planned and monitored. However, when reporting the project budget and schedule status, the report itself need only be at the level of the physical characterization task, not at the subtask and subsubtask level.

Figure 6.2 was taken from the same RFI QAPP as Figure 2.9. Note that the lowest level tasks in Figure 6.2 refer the reader to the data quality objective forms from the QAPP. These data quality objective forms provide information regarding matters such as the objective of the tests being performed, the procedures being utilized, and the level of quality assurance.

It is interesting to note in this example that the QAPP which originally contained these figures received written praise from the EPA Region IV, stating that the remediation service company that developed it demonstrated a thorough understanding of the data quality objective process. Such understanding was due, in part, to tying the data quality objective forms to the WBS. As a result of this letter, a great deal of customer trust was gained in the remediation service company's abilities. For a detailed discussion of data quality objectives, refer to "Data Quality Objectives for Remedial Response Activities" (U.S. Environmental Protection Agency, EPA/540/G-87/003 and 004, March 1987).

6.5 WORK BREAKDOWN STRUCTURE FUNCTIONS

The preceding sections provided some insights and examples regarding the WBS. This section discusses the functions of the WBS, with a detailed look at what the

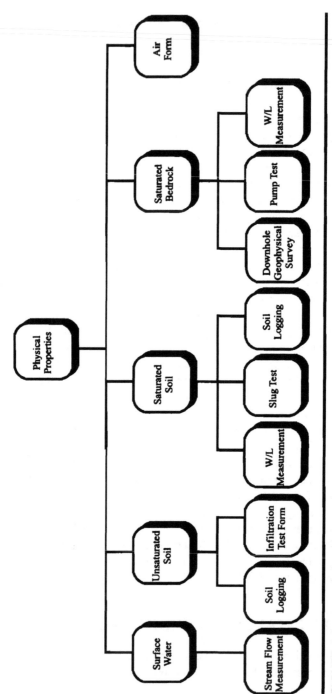

FIGURE 6.2 Expanded work breakdown structure for physical characterization.

WBS provides to the overall project management process. The WBS may be the single most important tool and technique in the project management process, since it provides (functions as) a common framework from which:[2]

- The total project can be described (as a summation of subdivided elements)
- Objectives can be linked
- Planning can be performed
- The level of quality control/performance can be determined (specifications)
- Schedules can be developed (in terms of duration of individual elements and network linking)
- Responsibility can be assigned
- Budgets can be established
- Information can be communicated
- Time, cost, and performance can be tracked (monitored)
- The project control system can be integrated with the company accounting system

6.5.1 Discussion of Functions

The central role of the WBS in project management is illustrated in Figure 6.3. Because of its multiple functions, developing the WBS is the next major step in the project planning and control cycle (see Chapter 1, Section 3.10, and Figure 3.13 for information regarding the project planning and control cycle), after determining the project objectives.

The total project/linked objectives. Sections 6.2.1 and 6.4 provided discussion and examples of how the WBS can function to present a project as a summation of subdivided elements and linked objectives. The following paragraphs discuss how the items in the above list of functions are accomplished by the WBS.

Planning. Planning is facilitated by the WBS because it forces the project team to think through the entire project by breaking it down into manageable work packages and organizing the work packages under logical headings.

Level of quality/specifications. The manageable work packages, as identified by the WBS, act as the starting point for determining the level of quality at which the work will be performed, as well as the specifications that will be met.

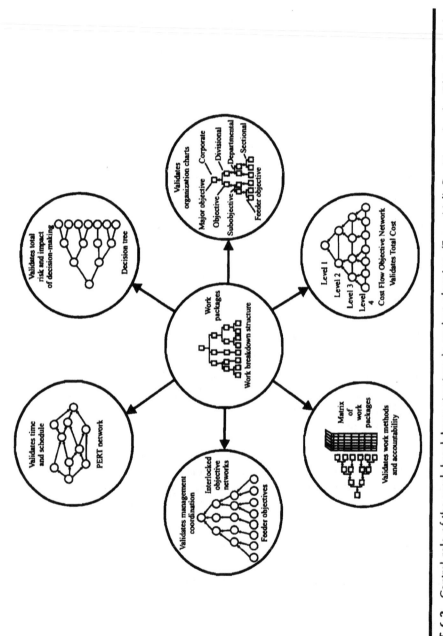

FIGURE 6.3   Central role of the work breakdown structure in project planning. (From Mali, P., *Managing by Objectives*, John Wiley & Sons, New York, 1972, 163. Reprinted by permission of John Wiley & Sons, Inc.)

Schedules, responsibility, and budgets. It is at the WBS work package level that schedules, the skill level of personnel, and budgets are all determined. Note that these items are greatly influenced by the required level of quality and specifications for each work package.

Communication. The WBS can be utilized to communicate, at the appropriate level of detail, to all project participants and stakeholders. For example, the customer may be interested in the cost and duration of the major tasks upon their completion. To provide information to the customer, the WBS may be "rolled up"; that is, information from the lower level tasks is summed up, forming the upper level report. At the other extreme, an individual responsible for a given work package can be provided with a detailed report regarding the schedule and budget for this detailed element.

Monitoring. Because all resources are assigned at the WBS work package level, actuals — in terms of cost, schedule, and percent complete — are recorded at this level.

System integration. As indicated above, the WBS acts as a common language (via the WBS code) for the sharing of information between a company's accounting system and the project planning program.

6.6 WORK BREAKDOWN STRUCTURE SPECIFICATIONS

The following specifications are offered to ensure that the WBS can serve its functions as discussed in Section 6.5. The WBS element must be:[2]

- ***Manageable***, in that specific responsibility can be assigned and durations/costs can be estimated
- ***Integratable***, so that the total project can be seen as the summation of the lower level elements
- ***Coded***, to permit the transfer of information among the planning, accounting, and reporting systems
- ***Measurable***, to offer quantifiable standards in terms of progress and performance

These specifications are the minimum requirements when reviewing a WBS for acceptability. The following pages offer additional guidelines to assist in development and review of a WBS. In many cases, the rules of thumb (provided in Section 6.8.2) may also sound like specifications. The difference is that, in certain situations, it may be desirable or even necessary to disregard some of the rules of thumb. However, the above specifications should never be disregarded.

6.7 TECHNICAL APPROACH

As discussed in Chapter 4, the technical approach is the basic strategy that will be utilized to reach the project objectives. The technical approach is conceptual in nature (i.e., all necessary details have not been fully worked out). Development of the technical approach involves identifying the major tasks that will be performed and conceptually how (level of effort, quality) these tasks will be performed.

The technical approach for a project is presented in Section 4 of the recommended proposal format (see Section 4.5). It should be possible to fully present the technical approach in two or three typed pages, even for the most complex projects.

Many project managers find it helpful to fully develop the technical approach before attempting to develop the WBS. This approach is recommended in most of the project management literature. However, in practice, many project managers develop the technical approach and the WBS in concurrent fashion. This is done by considering the project objectives and then discussing the potential primary tasks with the key project team members. While discussing these tasks, the team will begin to outline the WBS. Soon after or during development of the WBS, one of the team members is assigned the responsibility of documenting the technical approach. Either method is acceptable, and a project manager should use the approach with which he or she is most comfortable.

6.8 TECHNIQUES FOR WORK BREAKDOWN STRUCTURE DEVELOPMENT

This section provides:

- A general process for developing the WBS
- Rules of thumb for WBS development
- A methodology for team development of a WBS
- A discussion of WBS templates

6.8.1 General Process for Developing a Work Breakdown Structure

As previously stated, developing a WBS involves subdividing a project into smaller, more manageable elements. This subdividing must be continued until work packages can be defined in sufficient detail to support project planning, execution, and monitoring.

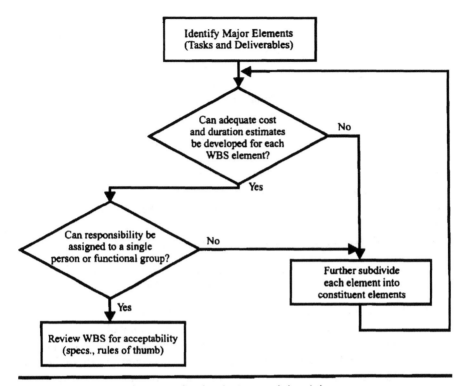

FIGURE 6.4   General process for developing work breakdown structures.

The general process for developing the WBS is described in the following steps and presented in Figure 6.4 (the process steps are taken from Reference 3):

1. Identify the major elements (tasks and deliverables) of the project.
2. For each major element, decide if adequate cost and duration estimates can be developed at this level of detail for each element. Also determine if responsibility can easily be assigned to a single person or functional group at this same level of detail.

   Note that the meaning of *adequate* may change over the course of the project. For example, subdividing a deliverable that will be produced far in the future may not be possible. If a deliverable cannot be divided into its constituent elements, then only conceptual (order-of-magnitude) estimates of cost and duration can be made for that deliverable. Such conceptual estimates for future deliverables may be considered adequate during the early phases of the project life cycle.

3. For those elements for which adequate durations and cost estimates cannot be made or responsibilities cannot be easily assigned, proceed to step 4. Otherwise, proceed to step 5.
4. Further subdivide each element into its constituent elements. Repeat steps 2 and 3 for each constituent element.
5. Review the WBS for acceptability by comparing it to the WBS specifications and rules of thumb.

Note that the WBS development process, as described above, permits different stages of the project to be broken down into different degrees of detail. For example, an earlier stage of a project (investigation) may be broken down into a significant amount of detail (subtask, subsubtask, etc.), while a later stage (design) may only be broken down into two or three significant tasks.

6.8.2 Work Breakdown Structure Rules of Thumb

The following rules of thumb can assist in the development of the WBS. They should also be used when reviewing the completed WBS.

- The names assigned to WBS elements should be easy to understand and convey meaning to all project participants and stakeholders.
- Include a task called project management. This will provide the project manager, and the core project team, with an account to bill against for project coordination, monitoring, and status reporting. (An exception to including this task would be a very small project, e.g., less than four weeks in duration.)
- Include a subtask entitled task management for each task that will be managed by a task manager.
- Think conceptually about the cost and duration of the lowest level elements during the development of the WBS, even though firm estimates of duration and cost will be made during later stages of the project planning process.
- The lowest level elements should account for from 5 to 10% of the total project budget. (An exception to this rule would be a smaller project that consists of only a few tasks.)
- The lowest level elements should not have a duration beyond two weeks. (An exception to this rule can be made for laboratory analysis and long-lead-time equipment.)
- Do not use the same task heading for activities performed by two separately responsible groups. (An exception to this rule can be made for

high-level summary tasks). For example, do not call a task *sampling* and *analysis* at the work package level. The sampling task may be performed by internal field personnel and the analysis task by a contracted laboratory. This calls for separate task names.

■ Be sure to include waste characterization and disposal tasks in the WBS whenever a remediation contractor will be responsible for these functions. If a remediation contractor is not responsible for these tasks, be sure to state this fact in the proposal assumptions and/or project contract.

■ Include a task in the WBS for the development of the project health and safety plan.

■ Include tasks in the WBS to update the health and safety plan as the project moves into new stages (e.g., from the investigation stage into pilot testing).

■ Include tasks in the WBS that will be performed by parties outside of the remediation service company (customer and regulatory review).

■ Include milestones in the WBS which represent the transfer of deliverables to parties other than remediation service company personnel (e.g., shipment of a report to a customer or regulatory agency).

■ Include milestones in the WBS which show that a major goal has been achieved (e.g., regulatory approval).

■ Include milestones in the WBS to signify the beginning or end of a project stage.

■ Break reports down into their component tasks, and show individuals responsible for writing, editing, and producing report components.

6.8.3 Group Development of the Work Breakdown Structure

A team approach is essential to creating a successful WBS for larger, more complex projects. This section provides a methodology for developing the WBS in a group setting.

Under ideal circumstances, all projects would be planned by the key project team members, working in a group setting. Under actual working conditions, as a rule of thumb, this method should be reserved for projects with a budget in excess of $250,000 and a duration greater than six months.

Section 5.3 of this book discussed methods of performing project planning when key team members are geographically distributed. Ideas discussed included the use of e-mail, voice mail system, and other communication tools.

6.8.3.1 Wall Planning Method for Developing the Work Breakdown Structure

The recommended method for developing the WBS in a group setting involves using a number of large self-adhesive notes and a very large blank wall. The group will utilize the notes to jointly construct the WBS on the planning wall. Although it may seem rather simplistic, developing the WBS using the wall method has proven to be highly effective in planning large projects performed by a major environmental service company.

The wall method allows team members to:

- See the complete WBS
- Voice suggestions and comments
- Quickly add, remove, and rearrange tasks as necessary

This method of collaboration is superior to gathering everyone around a computer screen. First and foremost, the complete WBS is usually too large to be viewed in its entirety on a monitor. Furthermore, project management software packages only automate the planning process. As with any computer program, they do not take the place of thinking through the process. Once the WBS has been completed on the wall, entering it into a computer is a simple task that can be completed by any member of the project planning team.

6.8.3.2 Materials

Large self-adhesive notes (4 × 6 inches) are recommended for the wall method, since they allow for large lettering that is visible at a distance. It is also helpful if the notes can be obtained in five or six colors, allowing one color for each major task.

6.8.3.3 Procedures

In using the wall method, each self-adhesive note represents a WBS element and must contain only one name. Experience has shown that it is best *not* to record WBS code numbers on the notes that contain the task names. Rather, it is better to place code numbers on separate notes, which are then mounted on the wall directly to the left of those notes that contain the appropriate task names. This allows the task names to be rearranged in any order without having to renumber them, while leaving the basic code structure in place.

The author's experience has further shown that when using the wall method for WBS development, it is better to use the indented outline format, as opposed to a tree diagram. (This is because the indented outline structure shows parent–

child relationships without having to draw lines on the wall.) Some individuals have found that a combination tree diagram/indented outline structure works well when developing a WBS using the wall method. This combination structure is presented in Figure 6.5, which shows each major task as a separate column heading, with an indented outline structure under each.

Items in addition to the WBS can be placed on the wall. For example, a large copy of the site map, groundwater gradient map, compound distribution map, etc. can be mounted, to help the team members visualize a site as they discuss the WBS.

Note that while the team members are discussing the WBS, they are likely to begin making a number of assumptions to facilitate their planning. The project manager should assign someone in the group the task of recording these assumptions as they are made. The recorded assumptions may later be inserted into the proposal, as a method of limiting the scope of work.

Additional points regarding the scope of work and quality specifications are likely to be raised during the development of the WBS. For example, the group might decide that ten wells will be installed during a well installation or that stainless-steel casing and screen will be used on all monitoring wells. Such decisions should be recorded to facilitate later development of the scope of work.

6.8.4 Work Breakdown Structure Templates

Ten WBS templates are provided in Appendix E. These templates will help initiate the development of a WBS.

WBS template 1 is a master template which shows the primary stages that are typical of a large-scale environmental remediation project. WBS templates 2 through 10 break down the project stages presented in the master template.

In referring to these templates, note that the names of the upper level elements (project stages) can be changed based on the environmental regulatory program (e.g., RCRA, CERCLA) under which a site is being managed. (For assistance with the names of project stages under various regulatory programs, see Table 13.2.)

6.9 SCOPE OF WORK BASED ON THE WORK BREAKDOWN STRUCTURE

In project management terminology, the scope of work is written text that describes the work to be accomplished. In other words, the scope of work describes

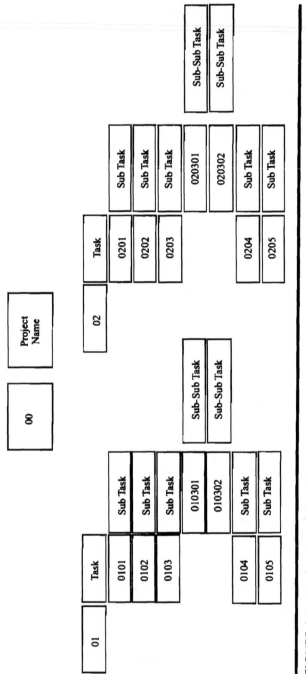

FIGURE 6.5 Combined work breakdown structure.

or represents the details (in terms of level of effort, quality of performance, resources supplied, materials used, etc.) of the work that is graphically displayed in the WBS.

The scope of work will appear in Section 5 of the recommended proposal format (see Section 4.5 of this book for a discussion of the recommended proposal format). Before a detailed scope of work can be developed, however, it is necessary to develop the WBS.

Once the WBS is complete, developing a written scope of work is a straightforward process. This is because the WBS serves to outline the scope of work. Using the WBS as an outline, the names of major tasks can become text headings for the various sections of the scope of work. In a similar fashion, the subtask names can become text subheadings and so on.

To facilitate writing the text, the project manager (or assigned team member) can use notes pertaining to project scope that were recorded during development of the WBS. To ensure that sufficient notes have been developed, it is recommended that the project team members review each of the WBS elements to ensure that items such as the following are discussed during development of the WBS and the decisions recorded:

- The number of wells
- The number and types of samples
- Pilot test procedures and duration
- Laboratory quality control procedures
- Health and safety procedures
- Waste-handling requirement

If the WBS was not developed in a group setting, then the individuals who were assigned the development of the WBS for a specific major task should also be responsible for developing the written scope of work for that task.

6.10 SPECIFICATIONS BASED ON THE WORK BREAKDOWN STRUCTURE

Specifications can be addressed as part of the proposal in the written scope of work or other documents, such as a quality assurance plan or detailed engineering specification list, can be referred to. Regardless of how all the specifications are presented, an effort must be made to identify them by considering the WBS at the work package level. For instance, the specifications can be related (but not limited) to items such as construction materials, analytical methodology, field quality assurance/quality control procedures, discharge limits, and sampling frequencies. As with the written scope of work, the specifications can be developed by the key project personnel working in a group setting or by the team member responsible for the planning of a specific task.

6.11 RISK MANAGEMENT CONSIDERATIONS

Methods of managing risk must always be sought throughout project planning and execution. (See Section 4.12 for the basics of risk management and definitions of terms such as risk factors, risk event, and expected value. In addition, Chapter 10 provides a detailed discussion of advanced project planning techniques which involve a rigorous analysis of risks.) Opportunities to identify project risk events — and develop response strategies — occur at nearly every stage in the project planning process. With each successive stage of planning, project risks can be increasingly understood and risk response strategies can become more detailed.

6.11.1 The Work Breakdown Structure and Risk Management

The WBS allows the project planning team to consider potential risk events at the work package level. Once possible risks have been identified, methods of managing those risks can be considered.

As discussed in Section 4.12.5, one of the most effective ways to manage project risks is by not taking responsibility for them in the first place. This is usually done by including a list of limiting assumptions in the project proposal. An example of a limiting assumption is that the customer will be responsible for the proper disposal of any waste generated during site investigation. Every attempt should be made to identify limiting assumptions associated with each WBS element, in order to mitigate risk.

6.11.2 The Scope of Work and Risk Management

Project risk is reduced when the written scope of work is highly detailed. This is because the more detailed the scope of work, the easier it is to identify and justify change orders. For example, if the scope of work calls for installing a monitoring well to a depth no greater than 20 feet, but field conditions indicate that the well should be installed to 25 feet, the case for a change order is easily made.

Although it is important to be specific when writing a scope of work, care should be taken not to make statements such as "XYZ company will fully determine the horizontal extent of soil impact." This type of statement may sound specific, but it is actually vague in that it does not describe the amount of work that will be performed to pursue this objective. A more specific (and less risky) scope statement would be: "XYZ will install a total of 40 soil gas-monitoring points in order to approximate the horizontal extent of soil impact."

6.12 EXERCISES

The exercises for this chapter are more extensive than those associated with the previous chapters. This is because the WBS represents the heart of the project plan. If it is developed properly, later project management tasks such as network diagramming, cost estimating, and monitoring will proceed more efficiently and accurately. It is also at the stage of WBS development that project strategies are confirmed, the technical approach is developed, objectives are verified, and limiting assumptions are made.

This chapter involves six exercises pertaining to the Megacorp site:

1. Development of the technical approach
2. Creation of a list of project objectives
3. Development of the WBS
4. Collection of a list of notes developed during the creation of the WBS
5. Formulation of a list of limiting assumptions
6. Identification of project risks

These exercises are in accordance with the project planning and control cycle presented in Figure 3.13. To effectively perform these exercises, items to have on hand include:

- The filled-in Request for Proposal (RFP) Summary Form
- A completed Stakeholder Management Summary Form
- A project organization chart
- Data provided in the RFP, including maps, tables, and drawings

A filled-in RFP Summary Form, Stakeholder Management Summary Form, and the project organization chart can be found in Appendix F. RFP information is provided in Appendix A.

During the project management training sessions out of which this book grew, the above exercises were performed in a group setting using the wall planning method. Obviously, the interaction, insights, and creative problem solving that came out of these sessions cannot be recreated within the confines of a book and while working alone. Therefore, answers to the exercises in the form of a written technical approach, objectives, assumptions, and project risks are provided below. In addition, a completed WBS for the project is included in Appendix G. The WBS appears in the third column from the left (indented outline format) of a report produced with the aid of Microsoft Project™. This report also presents the time-scaled network diagram (to be discussed in Chapter 7) and the project cost. To enhance and deepen your understanding of project management techniques, you may wish to attempt these exercises before looking at the answers. Example of notes that could be recorded when developing the WBS are also provided below. Notes such as these are extremely helpful when estimating project schedule and budget.

The development of a written scope of work is not included as an exercise in this chapter or elsewhere in the book. Such an exercise would be time consuming and require a lengthy write-up. The scope of work could be easily developed given the WBS and the notes provided below.

Considerations in Developing the Technical Approach

As stated in Section 6.7, the technical approach is the basic strategy that will be utilized to reach the project objectives. It is conceptual in nature, in that all necessary details have not been fully worked out. Development of the technical approach involves identifying the major tasks that will be performed and conceptually how (level of effort, quality) these tasks will be performed. The goal is to present the technical approach in two or three typed pages, even for the most complex projects.

According to the Megacorp site summary provided in Appendix A, Megacorp is looking for an environmental service company to:

1. Develop work plans for environmental investigation of the site, with a special focus on the former xylene underground storage tank (UST) area and the former disposal area
2. Investigate the site in accordance with the work plans
3. Pilot test an appropriate remedial action technology
4. Design the remedial action technology

Needless to say, these tasks must be included in the technical approach. However, additional insight into the project is obtained through completion of the Stakeholder Management Summary Form and the RFP Summary Forms. In particular, the Stakeholder Management Summary Form provides information regarding the expected behaviors of the various stakeholders. The RFP Summary Forms aid in the identification of project unknowns. In order for a project (or the proposal, for that matter) to be successful, it must strive to accomplish, as a minimum, two significant objectives in addition to those provided in the

RFP. First, it must strive to incorporate to the greatest degree possible the stakeholder response strategies as listed in the completed Stakeholder Management Summary Form (see Appendix F). The second is that it must, as cost-effectively as possible, convert significant unknowns into knowns (see the completed RFP Summary Form in Appendix F).

If the stakeholders are not managed effectively, they will surely work to negatively impact the project. The most effective technical approach gives a win to as many project stakeholders as possible. However, bear in mind that in many cases it is not possible to give a win to all stakeholders. This is especially true if giving a particular stakeholder a win would be to the detriment of the primary stakeholder (project sponsor) and the overall goals of the project. On the other hand, a win for the project sponsor that involves a definite threat to human health and the environment or breaking the law is not acceptable. Therefore, it should be understood that a technical approach that seeks to provide a win, within the stated constraints, to all parties has the best chance for successful project completion. Let's see how these concepts apply to the Megacorp project.

A review of the completed Stakeholder Management Summary Form indicates that Maureen Jones, Megacorp's environmental manager, and Paul Stanley, the corporate environmental lawyer, have conflicting goals. Maureen needs to accelerate the project and get a remediation system installed, in order to meet the challenge given her by upper level management (i.e., to have 15 remediation systems installed at sites that Megacorp plans to sell or where liability will be retained as part of the sales agreement). Paul Stanley, on the other hand, has the goal of trying to avoid commitments, slowing the project up, and waiting for it to fall in the realm of the CERCLA program. Is there a way to help both of these primary stakeholder achieve their goals? At first, it may seem impossible. This is where insights gained from filling out the RFP Summary Forms come into play.

The two contaminant source areas, the former xylene UST area and the former disposal area, are distinct in a number of ways. The first, which could easily be overlooked, is that the physical property itself has two different owners. Best Auto Parts owns the property where the former xylene area exists and Megacorp owns the property containing the former disposal area. The manner in which the contaminants of concern were released is significantly different in the two areas. The loss in the former UST area is by a tank failure; the loss in the former disposal area is the result of disposal practices. The primary contaminants of concern are different. In the former xylene UST area, the primary contaminant is xylene, while the primary contaminants of concern in the former disposal area appear to be polychlorinated biphenyls (PCB), solvents, and metals. Also, the information associated with the former xylene UST area is much more extensive and of higher quality than that associated with the former disposal area. Finally, it appears that xylene in the groundwater is moving away from the former xylene UST and perhaps off-site. Putting these elements together, it becomes apparent that groundwater impacts associated with the former xylene UST area should be addressed as soon as possible and that this area lends itself to the new voluntary cleanup program. On the other hand, based on groundwater sampling data in the former disposal area, there is not a significant threat of off-site impact at this time. Therefore, time is available to better organize a response for this area.

There is a way for both Maureen Jones and Paul Stanley to each achieve their objectives. This can be done by legally separating the site into two different sites; one site would include the former xylene UST area and the other would include the former disposal area. The first site could be known as the Best Auto xylene remediation site or simply the Best site (with Megacorp still responsible for cleanup). The other site could be known as the Megacorp former disposal area remediation site or simply the Megacorp disposal site. The Best site appears to be an ideal candidate for the voluntary cleanup program, while the Megacorp disposal site can proceed slowly, taking time to develop an investigation and remediation strategy in accordance with the CERCLA program. Separating the sites will involve legal and technical negotiations with both the state and federal EPA. This is best handled by Megacorp, with Paul and Maureen taking the lead.

Let's say that Enviroserv (the fictitious consulting firm responding to Megacorp's RFP, see Appendix A) identifies this strategy. Enviroserv has a better chance of winning the project if in selecting Enviroserv neither Maureen nor Paul must sacrifice their objectives. Perhaps Enviroserv would like to determine Maureen's and Paul's reaction to the concept of separating the site into two different sites. One way to make this determination would be to present the concept to Bill Preston (Megacorp's project engineer) first, since Bill would be Enviroserv's day-to-day contact with Megacorp should Enviroserv win the job. Assuming that Bill likes the concept, he can be the one to present it to Maureen and Paul, thereby involving him in the strategic planning and giving him an opportunity to impress Maureen and Paul. For the rest of this exercise, we will assume that this was done and that Maureen and Paul are in favor of the concept. We have satisfied three of the project stakeholders (Maureen, Paul, and Bill). Also note that the state's concerns over PCBs are directed more to the former disposal area. Therefore, if the site is broken into two separate sites, the Megacorp disposal site would be the one that has the potential to be placed on the National Priorities List. The approach of separating the site into two different sites may work toward satisfying the state's and the EPA's objectives as well.

We can now turn our focus to the Best site and potential remediation technologies. It is known that Tom Turbine (Megacorp's project manager), who has strong technical abilities, is of the opinion that high-vacuum dual-phase extraction (HVDPE) will work well at this site. This opinion is based on the heterogenous subsurface environment (glacial till) and the type of contaminant (xylene). During the Enviroserv team's planning session, it is discovered that Enviroserv's feasibility study engineer, design engineer, and project hydrogeologist also agree that HVDPE should work well at this site. Therefore this technology will be included in the project plan. The team will assume that the pilot test and the design will be for HVDPE. The project hydrogeologist also indicates that the two different water-bearing lenses (one at 8 to 10 feet, the other at 12 to 14 feet) are most likely in communication with one another. That is, groundwater and contaminants in the upper zone, although retarded to some degree by the clay layer between the two zones, can reach the lower zone. Therefore, it is best that the pilot test and the remediation system be designed to address both zones. The Enviroserv team may decide to install the HVDPE pilot test extraction well across both zones and to perform continuous split spoon soil sampling during the installation of the extraction well to make up for the fact that no soil logs are available from the previous investigation.

In order to satisfy the state's voluntary cleanup program, a feasibility study must be prepared to justify the technology that was selected for pilot testing and design. The Enviroserv team already has a strong feeling that such a study, if performed, would indicate that HVDPE would be the technology of choice. However, for the purposes of documentation in accordance with state requirements, Enviroserv will propose that a focused feasibility study be included in their scope of work. Finally, during the team planning session, the risk assessment manager might indicate that a risk-based corrective action approach could be used to establish site-specific cleanup goals for xylene (and perhaps PCBs in the groundwater).

Written Technical Approach for the Megacorp Site

Given the previously stated considerations, the technical approach for the project can be written. Recall that on even the most complex projects, the technical approach should not exceed more than two or three typed pages. Let's see if this can be done.

Technical approach. The technical approach proposed by Enviroserv involves having the Megacorp site legally defined and recognized by the state and the EPA as two separate sites. The first site, referred to as the Best site, includes the former UST area. The other site, the Megacorp disposal site, includes the former disposal area. Enviroserv believes that a strong case can be made for separating these sites, based on different property owners, different contaminants of concern, different release scenarios, and different threats to human health and the environment.

Separating the site into two different sites opens the door for including the Best site in the new voluntary cleanup program and accelerating site remediation, while at the same time permitting appropriate time to develop a high-quality response strategy for the former disposal area. It is Enviroserv's understanding that Megacorp would take the lead in negotiations with the state and the EPA and would establish a separate work order if it wants Enviroserv's assistance with these negotiations. The remainder of this technical approach will focus on the former xylene UST area (i.e., the Best site) in accordance with the recommendation to separate the site into two sites and manage the former disposal area at a later time, by means of a separate bidding process and work order.

The technical approach for the Best site involves six tasks:

■ Project work plans
■ Site investigation
■ Preliminary risk assessment
■ Focused feasibility study
■ Pilot test of HVDPE
■ Design of HVDPE system

The project work plans will consist of an investigation plan, health and safety plan, and quality assurance plan. The work plan for the pilot test will be developed as part of the overall pilot study task and after completion of the investigation, risk assessment, and focused feasibility study. The investigation will consist of a geoprobe soil-sampling program and sampling of all functioning groundwater wells (including those in the former disposal area, if functioning). The geoprobe soil-sampling program will determine the horizontal and vertical extent of xylene impact in the soil, which is not well documented by currently available data. Before groundwater sampling is performed, a well survey will be done to determine the conditions of the wells. All functioning wells will be gauged and developed. In addition, slug tests will be performed on two shallow wells and two deep wells for the purpose of determining hydraulic parameters.

Two rounds of a groundwater sampling will be performed. The first round will be on wells currently in place. Upon receipt of the sample results, locations will be selected for at most two new wells in the downgradient direction and near the expected edge of groundwater impact. Two soil samples will be collected from each new well bore during installation. The new wells are expected to be sampled within one month of the original groundwater sampling event. All collected samples will be analyzed for Target Analyte List (TAL) volatiles (method 8240), PCBs (method 8080), and TAL metals. All samples for metals analysis will be properly filtered in the field.

The preliminary risk assessment will be performed in accordance with ASTM Standard E-1739. The expected level of risk assessment is Tier 2. The focused feasibility study will look at a limited number of technologies and will follow guidelines established by the state.

The pilot test will begin with the development of a work plan based on all information available at the time. The test will be performed in accordance with Enviroserv's in-house guidelines. Tasks for pilot test execution will be initiated upon Megacorp's approval of the pilot test work plan. The pilot test results will be summarized in a written report which will provide the basis for design of the final remediation system. Upon Megacorp's acceptance of the pilot test report, detailed design will begin. The design documents will be of sufficient detail to permit release for bid.

In accordance with Megacorp's proactive approach, no activities will be delayed while awaiting regulatory approval. In addition, Megacorp, through its legal and environmental departments, will be responsible for all contact with the regulatory agencies.

List of Project Objectives

Provided below are objectives that could be included in the response to Megacorp's RFP. Note that the objectives as indicated in Section 4.5 usually are placed before the technical approach in a proposal. The objectives are categorized as long-term, short-term, and subobjectives, in keeping with the definitions provided in Section 4.8.3.

Long-Term Objective. The long-term objective for the Best site (former xylene UST area) is least-cost regulatory site closure.

Short-Term Objectives. The short-term objectives include:

1. The completion of an investigation which will provide sufficient information for a risk assessment and focused feasibility study
2. The establishment of risk-based cleanup goals acceptable to the state and technically achievable through the use of HVDPE
3. Proper documentation and support of HVDPE by way of a focused feasibility study
4. The completion of a pilot test that demonstrates the effectiveness of HVDPE and yields sufficient data for final system design
5. The development of detailed remediation system design documents sufficient for bid purposes and system construction

Subobjectives. The subobjectives include:

1. Approximation of the horizontal and vertical extent of soil impact
2. Approximation of the horizontal and vertical extent of groundwater impact
3. Estimation of the total mass of contaminants in the subsurface environment
4. Collection of physical and analytical data for fate and transport modeling
5. Determination of cleanup goals in soil and groundwater
6. Approximation of the radius of influence of the HVDPE
7. Estimation of the initial and long-term removal/destruction rate of xylene through the use of HVDPE
8. Selection of appropriate HVDPE system components and materials
9. Development of clear and detailed remediation system design drawings and specifications

The Work Breakdown Structure

Appendix G contains an example of a complete WBS that could have been created for the former xylene UST area by the Enviroserv team. This WBS was created with the aid of the:

1. WBS templates contained in Appendix E
2. Technical approach and objectives listed above
3. General process for WBS development (see Section 6.8.1)
4. WBS rules of thumb (see Section 6.8.2)

Example of Notes Recorded During Work Breakdown Structure Development

Listed below are notes the Enviroserv team might have recorded during the development of the WBS. The notes would then be incorporated into the written scope of work. These notes are not intended as a complete listing.

1. The geoprobe soil-sampling program will consist of no more than 20 borings.
2. The maximum depth for the geoprobe borings will be no more than 10 feet.
3. The geoprobe soil samples will be analyzed for TAL volatiles and PCBs.
4. Only two new wells will be installed during the extended investigation.
5. Groundwater samples will be collected from a maximum of 18 wells and analyzed for TAL volatiles, PCBs, and TAL metals.

6. All groundwater samples for metals analysis will be properly filtered in the field.
7. A quality assurance plan will be developed for the extended investigation.
8. Megacorp will be responsible for the disposal of waste soil and groundwater generated during the extended investigation. Enviroserv will be responsible for placing the materials in drums, properly labeling the drums, and collecting characterization samples.
9. The HVDPE pilot test will be conducted over a two-day period.
10. Groundwater obtained during the HVDPE pilot test will be collected in a 500-gallon storage tank. Enviroserv will collect waste characterization samples. Megacorp will arrange for a disposal company to pump out the tank.
11. The HVDPE pilot test extraction well and monitoring points will be constructed in such a manner that they will be incorporated into the final design.
12. The extraction well will be installed with schedule 80 continuous wrap well screen.

Limiting Assumptions

Provided below are examples of assumptions that might be identified to limit the project scope of work for the former UST area project (Best site).

1. Megacorp will perform negotiations with the state and the EPA regarding separating the site into two distinct sites.
2. The negotiations with the state and the EPA will be successful.
3. The former UST area will be managed under the state's new voluntary cleanup program.
4. Significant contaminants other than xylene and PCBs will not be found in the former UST area.
5. Field activities will be initiated upon Megacorp's approval of the investigative, health and safety, and quality assurance plans. Field activities will not be delayed while waiting for state approval of work plans.
6. Megacorp will not require substantial revision to any of the work plans or reports.
7. The expected outcome of the focused feasibility study is the recommendation for HVDPE.
8. At least ten of the currently installed wells are in functioning condition. No more than three wells will be repaired. The repairs will be minor in nature (i.e., surface casing, well box, etc.). No currently installed well will be redrilled.

9. The pilot test technology will be HVDPE.
10. The pilot test will yield positive results; if it does not, additional pilot testing and design costs will be incurred.
11. Site work will not be delayed by the activities of Best Auto Parts.
12. Megacorp will approve all boring locations and will determine the locations of all subsurface utilities, obstructions, etc. that may be present at the site.
13. An air discharge permit will not be required for the HVDPE pilot test (discharges below permitted levels).
14. The electrical transformer at the site did not leak and PCBs are not expected to be a significant contaminant in the former xylene UST area.

Project Risks

The largest risk associated with this project is the fact that it is being performed under a firm fixed price contract. The risks have been significantly reduced by the use of limiting assumptions (e.g., at least ten wells are in functioning condition, no subsurface obstructions, etc.) and a detailed scope of work (based on the notes collected during the WBS development). However, examples of project risks that may occur during project execution include:

1. The EPA may issue a consent order for the site before it can be placed in the voluntary cleanup program.
2. Best Auto Parts could create significant and repeated delays in site activities.
3. The monitoring wells could be significantly silted up, requiring extra time for development.
4. There could be a shortage of equipment needed for the HVDPE pilot test, such as the liquid ring pump. Therefore, costs for HVDPE equipment could increase.
5. An unidentified subsurface obstruction, such as a sanitary sewer, could be penetrated during drilling operations.
6. A significant Enviroserv team member, such as the design engineer, could leave the company halfway through completion of his or her assigned task.

Final Note

The completion of these exercises marks a significant stage in the completion of the project plan. The project strategy has been determined, objectives have been verified, and the tasks that must be performed have been identified. The next two chapters deal with scheduling and budgeting for carrying out these tasks, which are the next steps along the project planning and control cycle.

REFERENCES

1. Project Management Institute Standards Committee, *A Guide to the Project Management Body of Knowledge (PMBOK), Exposure Draft*, Project Management Institute, Upper Darby, PA, 1994, 62.
2. Kerzner, H., *Project Management: A Systems Approach to Planning, Scheduling, and Controlling*, 2nd ed., Van Nostrand Reinhold, New York, 1984, 553.
3. Project Management Institute Standards Committee, *A Guide to the Project Management Body of Knowledge (PMBOK), Exposure Draft*, Project Management Institute, Upper Darby, PA, 1994, 22.

Network Diagramming/Scheduling

7

Preparation of the network diagram and the master schedule represents steps four and five in the project planning control cycle (see Figure 3.13 and Section 3.10 for a discussion of the project planning and control cycle). Together, the project network diagram and master schedule provide the basis for time management (one of the nine project management functions). This chapter provides information, tools, and techniques to facilitate the development of the network diagram and project schedule. In addition, this chapter explains the process of project time management.

Time management is often the most challenging of the nine project management functions; perhaps this is because time is inflexible. The importance of time management is widely recognized, as emphasized by the following quote:

> Time is...a unique resource. Of the other major resources, money is actually quite plentiful....People, one can hire, though one can rarely hire enough good people. But one cannot rent, hire, buy or otherwise obtain more time.
>
> Peter F. Drucker, *The Effective Executive*

7.1 TIME: A RESOURCE OR A CONSTRAINT?

Although the above quote indicates that Mr. Drucker understands the importance of time management, it is interesting that he chose to refer to time as a resource. Traditional management views time as a resource, while project management does not. (Recall that the six primary resources available to the project manager are money, personnel, equipment, materials, information/technology, and facilities; see Section 3.6.)

Within the project management literature, there has been some controversy over the years as to whether or not time is indeed a resource. In the early project management literature (1950s to 1960s), some project managers chose to think of time as a resource because it must be managed and because, in a project context, it is possible (although risky) to obtain more time, in the form of customer-approved project extensions. More recently, many project managers have chosen to reserve the term resource for those items that one can rent, hire, or buy in order to meet the goals and objectives of a project. These project managers view time as a constraint, within which resources must be managed. The project management literature of the late 1980s and early 1990s favors the position that time is indeed a constraint rather than a resource.

7.2 TIME MANAGEMENT VERSUS CRISIS MANAGEMENT

Initially, the need for time management may seem obvious, given the customer-imposed deadline for project/task completion. However, the more important — and not so obvious — reason to study time management techniques is to learn how to prevent crisis management. Crisis management is often the opposite of time management. It usually occurs when we discover (rather than anticipate) that we have looming deadlines and the labor hours required by the project far exceed the hours available from project personnel.

Much of the project manager's job is concerned with managing resources — within the constraints of time. If a management style provides little understanding of what resources will be needed, and by what time, the result will always be a crisis management scenario.

7.3 CRISIS MANAGEMENT AND ENVIRONMENTAL REMEDIATION

A characteristic that seems common to most environmental remediation projects is the amount of crisis management that occurs (due to project requirements rather than toxic or hazardous situations). The September 1994 issue of the *Project Earth Quarterly* (published by the Project Management Institute) contained an article entitled "Environmental Projects in Need of Better Project Management."[1] The authors of this article interviewed approximately 90 project managers from three different organizations involved in environmental remediation projects. Topping the list of project managers' concerns (out of 33 possibilities) was avoiding crisis management and fire fighting. The way to address this major concern is to study and implement the time management techniques (network diagramming/scheduling) discussed in this chapter.

7.4 WHAT IS SO DIFFICULT ABOUT SCHEDULING?

To those new to project management, the development of a schedule may seem a straightforward process, involving simple decisions, regarding which activities we would like to perform first. Actually, scheduling is quite complex — although it need not be difficult. Keep in mind that in scheduling, the order of activities should not depend solely on personal preferences, nor should scheduling be approached lightly.

Accurate project scheduling requires an understanding of:

- The interdependencies among project activities and events (network diagramming)
- The duration of the various activities that comprise the network (duration estimating)
- The availability of resources (resource leveling)
- Mathematical methods for analyzing these interdependencies and availabilities (critical path method and program evaluation and review technique)

7.5 CHAPTER TOPICS

Modern project management had its beginning in the search for techniques to graphically represent and mathematically analyze activity interdependencies, durations, and resource availabilities in order to improve project scheduling. The techniques that resulted from this search are the focus of this chapter.

The primary topics discussed in this chapter include:

- A brief history of network-based scheduling techniques
- A formal definition of the project schedule
- Project time management
- Activity sequencing
- Duration estimating
- Network calculations

- Guides to successful scheduling
- A scheduling case study exercise

7.6 HISTORY OF NETWORK SCHEDULING

Until the advent of network diagramming and critical path analysis techniques, there were no generally accepted formal procedures to aid in the development of project schedules. In the past, most project managers had their own methods of scheduling, which may have involved some use of the bar chart format originally developed by Henry Gantt (in approximately 1900).[2] Although the bar chart is still a useful tool in project management, it is inadequate as a means of describing the complex interrelationships among project activities.

It was not until 1957–58 that a more formal approach toward project scheduling occurred. Within this time frame, several techniques were developed concurrently, but independently.[2] The technique known as the critical path method (CPM) was developed in connection with a very large project undertaken by the DuPont Corporation.[3] The objective that led to the development of CPM was to determine the optimum duration (based on minimum cost) for a project whose individual task durations were primarily deterministic variables (in other words, the task durations could be estimated with a relatively high degree of accuracy).

A development similar to CPM occurred in Great Britain, where the problems of overhauling an electricity-generating plant were being studied.[4] In this case, the objective which led to the CPM-like technique was to determine the "longest irreducible sequence of events."

Yet another approach to project scheduling, called the program evaluation and review technique (PERT), was developed in conjunction with the U.S. Polaris weapons system. The objective here, which led to the development of PERT, was to develop an improved method of scheduling an extremely large and complicated research and development project, where many of the required activities were state of the art or beyond.[5] Scheduling this kind of research and development project is obviously very difficult, especially because activity durations are random variables that involve considerable variance (note that such variables are often referred to as stochastic or probabilistic).

Although the above techniques were developed independently, they all are based on the important concept known as the *network representation* (diagram) of the project plan. Later in this chapter, modern methods of developing network diagrams (precedent diagram, activity-on-arrow diagram), as well as methods of analyzing the diagrams (CPM, PERT), are discussed in detail.

7.7 SCHEDULE DEFINITION

The Project Management Institute's current definition of the project schedule is as follows:

> **Project schedule:** The planned dates for performing project activities and meeting project milestones.[7]

Although this definition is short and simple, some discussion of its terms is in order.

Activities, as used in the above definition, refer to the project work packages (i.e., the lowest level elements in the work breakdown structure). As noted in Chapter 6, all project work is performed at the work package level (i.e., all activity occurs here), and, in fact, all time and resource consumption occurs at this level as well. The upper elements in the work breakdown structure act merely as summary headings, to categorize the work packages.

Note that although the project schedule also indicates the planned dates for the performance of the summary tasks, such dates are determined by scheduling the lowest level elements (i.e., the project work packages or "activities").

Milestones refer to key events in a project, with an event defined as the starting point (or ending point) of a group of activities. Keep in mind that an event (or milestone):

- Is a point in time, without duration
- Is said to be complete when all activities leading to it are complete
- Does not, in and of itself, consume resources

A final point about a project schedule is to recognize how it differs from a network diagram. A network diagram is a schematic display of the logical relationships among activities and events. A project schedule, on the other hand, is a time schedule based on a mathematical analysis of the network diagram.

7.8 PROJECT TIME MANAGEMENT

Effective time management includes a number of processes required to ensure timely completion of a project. The five steps involved in project time management are introduced below and presented in Figure 7.1.

Activity (task) definition. The first process in project time management, activity definition, involves identifying the specific activities that must be per-

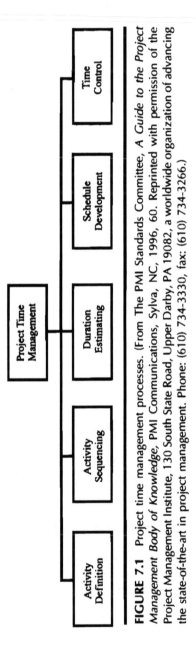

FIGURE 7.1 Project time management processes. (From The PMI Standards Committee, *A Guide to the Project Management Body of Knowledge,* PMI Communications, Sylva, NC, 1996, 60. Reprinted with permission of the Project Management Institute, 130 South State Road, Upper Darby, PA 19082, a worldwide organization of advancing the state-of-the-art in project management. Phone: (610) 734-3330, fax: (610) 734-3266.)

formed in order to reach the project objectives.[7] Activity definition is accomplished through the development of the project work breakdown structure, the written scope of work, and the project specifications (discussed in detail in Chapter 6).

Activity (task) sequencing. The second process in project time management, activity sequencing, involves identifying and documenting the dependencies that occur among all the project activities and events (i.e., developing the network diagram words). (See Section 7.9 for a detailed discussion of activity sequencing.)

Duration estimating. The third process in project time management, duration estimating, involves assessing the number of work periods (e.g., hours, days, weeks) required to complete the individual activities. (See Section 7.10 for more on duration estimating.)

Schedule development. The fourth process in project time management, schedule development, involves analyzing activity sequences, activity durations, and resource requirements to create the overall project schedule.

Time control. The fifth process in project time management, time control, involves maintaining the project schedule, through monitoring, corrective action, and schedule updates. Note that time control is a function of project execution and monitoring, as opposed to planning, which is the primary focus of this book. (See Chapter 9 for an introduction to project time control.)

7.9 ACTIVITY SEQUENCING

Activity sequencing deals with the connections and interdependencies that exist within a project. The activity sequencing process is facilitated by first fully defining the activities and events; therefore, it is helpful to have on hand the items discussed in Chapter 6. Such tools include the completed work breakdown structure, written scope of work, identified specifications, and a written list of assumptions (developed during construction of the work breakdown structure).

Activity sequencing relies on an understanding of:

- Dependency categories
- Network diagramming methods
- Network logic connectors (dependency types)
- The general process for developing the network diagram

7.9.1 Activity Sequencing Dependency Categories

Dependency is the timing relationship that exits between two tasks (or events). The earlier task in a dependency relationship is referred to as the predecessor, and the latter is known as the successor.

In order to develop an accurate network diagram, one must have an understanding of the various types of dependency relationships and be able to apply this understanding to the project at hand.

Note that there are two major categories of dependency relationships: mandatory and discretionary.[7]

7.9.1.1 Mandatory Dependencies

Mandatory dependencies are often referred to as technical or physical constraints. This type of dependency occurs naturally as a result of the work being performed or the technology being employed. For example, during an environmental investigation, a mandatory dependency might be that the groundwater samples cannot be collected until the monitoring wells have been installed. Another example of a mandatory dependency might be that the structure for the equipment building cannot be constructed until the foundation has been built. Because mandatory dependencies must be adhered to in the context of a network diagram, they are often referred to as *hard logic*.[7]

7.9.1.2 Discretionary Dependencies

The other major category of dependency relationships, discretionary dependencies, includes those which are not subject to mandate — yet which are also important to the network development process.

Often these discretionary dependencies are built on *soft logic*,[7] such as the knowledge of "best practices." An example in environmental remediation would be a case where the team may know from past experience that during the construction of a high-vacuum dual-phase extraction system, it is best to install all of the extraction points before installing the equipment compound. Although it is not mandatory that these tasks be performed in this order, such an approach may reduce construction site congestion.

A discretionary dependency can also be derived from *preferential logic*,[7] such as a start date specified by a project stakeholder. For example, the customer may wish to install and start up a remediation system prior to completion of the investigation and final regulatory approval of the technology type. Such dependency relationships are still discretionary in nature since the customer has the option to alter this approach.

7.9.2 Network Diagramming Methods

The logic used in activity sequencing, such as the dependencies discussed above, will eventually be structured within the network diagram. There are essentially three network diagramming methods (formats) currently used within the field of project management:

- Precedent diagram (also known as the activity-on-node diagram)
- Activity-on-arrow diagram
- Time-scaled network diagram

Figure 7.2 presents a comparison of these three diagramming formats, based on a pilot study stage of a remediation project. (Note that the number of tasks within the pilot study has been reduced and simplified for the purposes of this example.)

7.9.2.1 Precedent Diagram

In the precedent diagram method, the activity names are written in the diagram boxes and arrows are used to indicate how the activities are related (logical connections).

The precedent diagramming approach is the top choice of many project managers and is readily available on project management software. Most individuals find the precedent diagram natural and easy to work with. Perhaps the major advantage of the precedent diagram is that a variety of logic connectors may be utilized, as opposed to the finish-to-start limitation imposed by the activity-on-arrow method. (Network logic connectors are discussed in more detail in Section 7.9.3.)

7.9.2.2 Activity-on-Arrow Diagram

The activity-on-arrow diagram is just the reverse of the precedent diagram. As indicated by the name, the activity-on-arrow diagram places the activity names above the arrows that construct the diagram, and logic constraints are indicated by event nodes. The event nodes represent points in time when all activities leading into an event have been completed.

The activity-on-arrow diagram permits only finish-to-start logic connections; therefore, dummy activities must be inserted into the diagram to properly indicate all network logic. The activity between events 5 and 3 in Figure 7.2b is an example of a dummy activity. Over the past few years, the activity-on-arrow method has rapidly fallen out of favor due to its complexity and difficulty to use. Therefore, this method will not be discussed further.

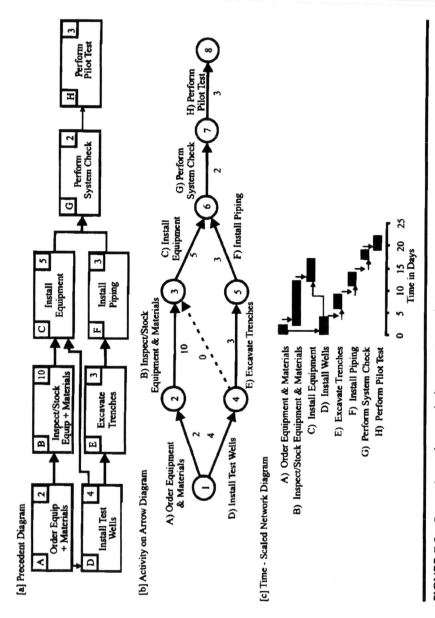

FIGURE 7.2 Comparison of network diagramming techniques.

7.9.2.3   Time-Scaled Network Diagram

A third type of network diagram method, the time-scaled network diagram, is a format which presents the project logic and the activity durations. In effect, it is a blend of a network diagram and bar chart, thereby correcting the traditional weaknesses of both (i.e., traditional network diagrams did not present calendar durations, and bar charts did not show the logical links between tasks).

The time-scaled network diagram can be presented in either a precedent or activity-on-arrow format. Figure 7.2c is a time-scaled diagram presented in the precedent format.

Most project managers have found that, in practice, it is best to develop the activity sequence first and then determine the *duration* of the various activities. Note, however, that activity sequencing and duration estimating are somewhat independent functions; the order in which to perform these operations can be reversed, if desired, by a given project manager.

Deciding in what order to perform activity and duration estimating was more crucial when networks were developed by hand. Many of the newest project management programs permit the display of the time-scaled network diagram on a computer screen while the project is being developed, thereby facilitating activity sequencing and duration estimating simultaneously.

The time-scaled network diagram is rapidly gaining favor, since it has become available on most PC-based project management software packages. Nearly all of these software packages present the time-scaled network diagram in precedent format. In other words, all of the logic connectors used in the precedent diagramming approach are available.

In the remainder of this chapter, the generic term network diagram will refer to the precedent diagramming method — with the understanding that the time-scaled network can be easily constructed once the activity durations have been determined.

7.9.3   Network Logic Connectors

As we look at activity sequencing in this chapter, we have covered dependency categories and network diagramming methods. Our next consideration is network logic connections. There are four primary types of network logic connectors as presented in Figure 7.3 and discussed below. It should be noted that any of these logic connections may fall within the category of mandatory or discretionary dependencies.

- ■ *Finish to start*: The "from" activity must be finished before the "to" activity can start.
- ■ *Start to start*: Essentially, both tasks start at the same time.

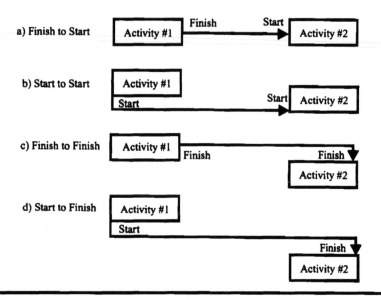

FIGURE 7.3 Types of network logic connections.

- *Finish to finish:* The "from" activity must be finished before the "to" activity can be finished.
- *Start to finish:* The "from" activity must start before the "to" activity can finish.

The **finish-to-start logic connector** is the most common. Finish-to-start dependency types usually fall within the category of mandatory dependencies. If the customer desires, the overall project schedule can be extended to the maximum length possible by using all finish-to-start dependencies.

The **start-to-start logic connector** is the second most common type of dependency used in network scheduling. These start-to-start connectors cause tasks to be performed in parallel, shortening the overall schedule. Start-to-start dependencies can usually be utilized anywhere mandatory dependencies do not exist. Under the conditions of a competitive request for proposal, one may wish to maximize the use of start-to-start logic connectors.

The **finish-to-finish logic connector** is used occasionally. The most common use of a finish-to-finish relationship is to indicate that a task management subtask will not be complete until all other subtasks have been completed.

The **start-to-finish logic connector** is permitted by nearly all project management software packages, yet for the most part is never used. By simply changing the order of the tasks in the network diagram, this logic connector can be replaced with a finish-to-start logic connector.

7.9.3.1 Partial Dependencies (Lead and Lag Relationships)

The use of partial dependency relationships is common in most projects. Partial dependencies are additional constraints that may be placed on any of the four logic connectors discussed in the previous section and are used primarily to create leads or lags in activity sequencing. Figure 7.4 presents examples of lead and lag sequencing. As indicated in Figure 7.4A, a lead relationship is one where it is necessary for task number 1 (the predecessor) to be have been *initiated* for a set period of time or to have achieved a certain percentage of completion before task number 2 (the successor) can start. Figure 7.4C indicates that a lag relationship occurs where it is necessary for task number 1 (the predecessor) to have been *completed* for a set period of time before beginning task number 2 (the successor).

An example of a lead relationship in an environmental investigation would be the case where sampling activities (i.e., the predecessor task) may proceed for two or three days prior to the initiation of laboratory analysis (i.e., the successor). This two- or three-day period could represent, for instance, the time it takes to collect a sufficient number of samples, prepare transport coolers, and transport the coolers to the laboratory.

An example of a lag relationship that occurs during an environmental investigation is the case where it is desirable to wait one week after developing newly installed site wells (i.e., the predecessor task) prior to starting sampling activities (i.e., the successor).

Although, in theory, partial dependency constraints can be placed on any of the four logic connectors, lead and lag relationships are used primarily with start-to-start and finish-to-start logic types. A partial dependency constraint is placed on start-to-start logic connectors to create activity *leads* (start task 1, wait several days, start task 2), whereas partial dependencies are used with finish-to-start logic connectors to create activity *lags* (finish task 1, wait one week, start task 2).

7.9.3.2 Multiple Logic Connectors

The various activities and events within the network diagram may have multiple dependencies as well. Note that in Figure 7.2a, activities C (install equipment) and G (perform system check) provide examples of activities with multiple dependencies.

Care must be taken to identify and include all mandatory multiple dependencies in the project schedule. They have a significant impact on how the activities and events may be sequenced, which will become more apparent during the discussion of schedule development (Section 7.11).

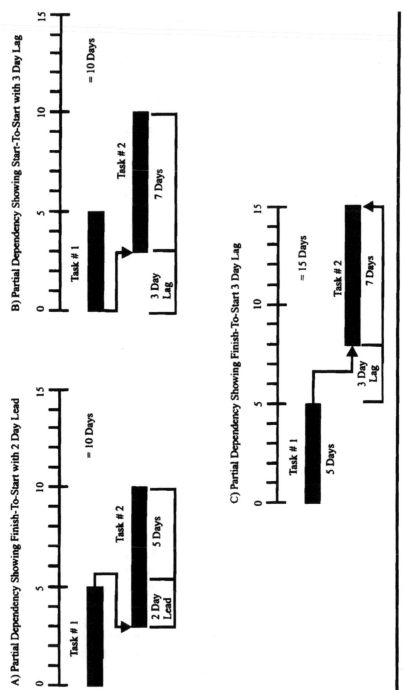

FIGURE 7.4 Lead–lag sequencing using partial dependencies. (From Houshmand, Z., Livezey, R., and Shea, C., *Time Line Users Manual*, Version 4.0, Time Line Solutions Corporation, Novato, CA, 1990, p. 7-5. With permission.)

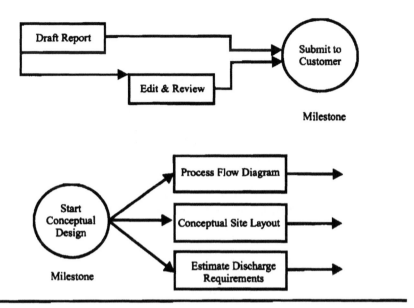

FIGURE 7.5   Examples of multiple dependencies.

Multiple dependencies tend to occur at project milestones (i.e., key project events), although they can occur at other locations within the network as well. Some examples of project milestones in the environmental remediation industry are

- The transfer of deliverables to parties outside the environmental service company
- Achievement of a major goal (regulatory approval)
- The beginning or end of a major project stage (risk assessment)

Figure 7.5 presents two examples of multiple dependencies occurring at project milestones.

7.9.4   General Process for Developing the Network Diagram

Based on the above theories of activity sequencing, one can assemble a practical network diagram. The following steps represent the general process for constructing the network diagram:

1. Review the work breakdown structure, giving particular attention to activities (work packages) and milestones.

2. *For each activity and milestone*, ask the following questions:[8]
   - What element immediately follows?
   - What element immediately precedes?
   - What elements can be performed concurrently?
3. Using a project management software package, link all the activities and milestones together, to create the network diagram.
4. After the first pass, review the network diagram to ensure that:
   - All mandatory dependencies have been included
   - All necessary multiple dependencies have been included
   - There are no errors in logic (The primary errors in logic include looping and dangling tasks, discussed below.)
5. Review the network diagram for acceptability by considering the rules of thumb (see Section 7.9.4.2).

7.9.4.1   Common Errors in Logic

Presented below are the two most common errors in network logic: looping and dangling tasks.

Looping.   If a loop is drawn in the network, it is clearly a logic error.[9] Once into the loop, one can never get out — and the project will go on forever.[9] While a loop may seem an accurate representation for some cases, projects should never be planned that way![9] In Figure 7.6, the loop is easy to see. Note, however, that in a larger network that involves many activities and arrows, such loops could go unnoticed.[9] Fortunately, most project management software packages will detect this logic error.

Dangling tasks.   A dangling task is one that has not been connected back into the network, as indicated in Figure 7.7. Such a task must be connected back into

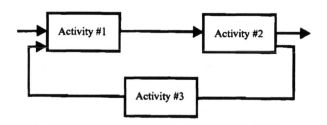

FIGURE 7.6   Network loop. (From Mulvaney, J.E., *Analysis Bar Charting: A Simplified Critical Path Analysis Technique,* Management Planning and Control Systems, Washington, D.C., 1989, 12. With permission.)

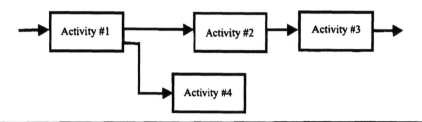

FIGURE 7.7 Dangling task. (From Mulvaney, J.E., *Analysis Bar Charting: A Simplified Critical Path Analysis Technique,* Management Planning and Control Systems, Washington, D.C., 1989, 12. With permission.)

the project (even if only to finish a milestone), or the project will have two ends.[9] Some project management software packages will recognize this type of logic error, while others will not.

7.9.4.2 Network Diagramming Rules of Thumb

The following rules of thumb can assist in the development of a network diagram. They should also be used when reviewing the completed network diagram. Note that these rules of thumb are guidelines. Depending on the specific project, it may be necessary (even desirable) to take a different approach.

- Discretionary dependencies should be kept to a bare minimum. Such dependencies add complexity to the network logic and can extend the project duration.
- Maximize concurrency by using start-to-start links whenever possible, where their use is not prevented by mandatory dependencies. Note that indicating that two tasks can start concurrently does not mean that they *must* start concurrently. (The actual start date of a task will be determined by the mathematical analysis of the network logic, discussed in Section 7.11.)
- Be sure to link only the lowest level elements in the work breakdown structure (activities and milestones). Some project management software packages now permit linking of summary tasks. This forces unnecessary logic and will limit the start date options when mathematically evaluating the network logic.
- Avoid including preferential logic, such as customer-mandated start dates, in the network diagram. One should initially prepare the network diagram without including preferential logic and then, based upon mathematical analysis, see if the customer-imposed date can be met without forcing

such logic. If the required date cannot be achieved, then adjustments can be made to the network logic or resource assignments.

■ Have the task managers develop (or review) the logical links for the tasks for which they are responsible. Be certain that each task manager understands the networking concepts discussed in this chapter.

7.10   DURATION ESTIMATING

So far, we have covered the first two concerns of time management, activity definition and activity sequencing (see Figure 7.1). The next process in time management, duration estimating, is discussed below.

Duration estimating involves assessing the number of work periods (hours, days, weeks) required to complete each individual activity. Note that duration estimating involves much more uncertainty than activity sequencing. With activity sequencing, one must either adhere to mandatory dependencies or choose discretionary dependencies. In either case, we can be relatively certain that the sequence of activities will be performed in the planned order. With duration estimating, however, we are dealing with time, which always involves uncertainty.

In an ideal world, we would have at our disposal a large volume of historical data upon which to base these estimates.[16] Although such a database has yet to be developed, as more of the techniques discussed in this book are put into practice, such a tool may begin to evolve. (Keep in mind, however, that even if such a database does evolve, many activities will be project specific and therefore cannot be identical to historical data.)

7.10.1   Duration Estimating Guidelines

The following guidelines are offered to assist in the duration estimating process and to improve estimating accuracy. Note that these guidelines have been developed assuming that resources are being considered conceptually, but not actually assigned at this time. (See Chapter 8 regarding assigning resources to tasks.)

Obviously, the labor resources assigned to a task can affect its overall duration. It is common to estimate task duration first by thinking conceptually about the level of resource assignments. Once resources are actually assigned, the activity durations can be further refined and adjusted as necessary.

The following guidelines can be used for duration estimating:

■ Whenever possible, make use of in-house experience. Seek out individuals who may have performed similar activities in the past.

■ Have the people who will be responsible for performing a given task provide duration estimates. This does not mean talking only to task managers, but also to technicians, field engineers, and field geologists. These individuals have a tremendous amount of knowledge that should not be overlooked.

■ Make all time estimates realistic. It is best to consider the most likely amount of time needed, without an optimistic or pessimistic view.

■ Assume that the work will be performed with a normal level of labor and equipment. Although completion of tasks can often be accelerated by adding more people and/or equipment, it is best not to assume an exaggerated level of effort. An exception would be where an analysis of the network indicates that the customer's required finish date cannot be met with a normal level of effort.

■ Assume a normal workweek and a normal workday.

■ Use consistent time units for all activity durations. (It is usually best to estimate activity durations in days and resource allocations in hours.)

■ Be willing to break activities into smaller units (work packages) if there seems to be great difficulty in estimating a particular activity. (If this is done, the work breakdown structure and network diagram will require updating.)

■ Do not assume that an office worker will have eight hours of uninterrupted time to work on a task. Office workers are constantly interrupted by phone calls, other office personnel, meetings, etc. Assume that an office worker is 70% productive — and adjust task durations accordingly.

■ Because reports require revisions, be sure to include editing time (beyond your original estimate) for making the necessary changes.

7.11 SCHEDULE DEVELOPMENT

In this section, the fourth process in time management, schedule development, is discussed. Schedule development can begin once all the activities (and milestones) are linked in a logical sequence and the duration of all activities has been estimated.

Schedule development can be seen as determining the start and finish dates for all project activities. Note that arriving at a final schedule often requires a number of iterations through the processes utilized to develop the schedule.

The commonly used methods for developing the project schedule include:

■ CPM
■ PERT
■ Resource-leveling heuristics

Each of these techniques will be discussed in the following sections. It should be noted that algorithms for performing these analyses are included in most project management software packages (with the possible exception of PERT). Therefore, it is not necessary to understand these algorithms. However, many people have reported difficulty in learning project management software or have stated that these programs are not user friendly. This difficulty most likely stems from their not understanding what the software is attempting to do through the processing of these algorithms.

The CPM technique will be discussed in the greatest amount of detail because it is the most useful application for determining activity start and finish dates. The PERT technique will be discussed in the least amount of detail, because it is not commonly used (although it is expected that more project management software programs will offer this feature in the future).

The CPM often produces a preliminary schedule that requires more resources than are available at certain times. When this happens, the schedule must be adjusted to "level out" overscheduled resources. Resource leveling often results in a project duration that is longer than the preliminary schedule. Resource leveling will be explained in a moderate level of detail so environmental project managers can begin to make use of this technique. Incorporation of this technique into project planning could greatly reduce the amount of crisis management.

Before any of the above analytical methods can be employed, it is important to set the project calendars within the computer software. Project and resource calendars identify periods when work is allowed. Project calendars affect all resources. These calendars account for weekends and planned company holidays. Resource calendars affect a specific resource or category of resources (e.g., a team member on vacation, field personnel who generally work ten hours per day).

7.11.1 The Critical Path Method

As discussed earlier in the chapter, development of the network diagram places all the activities in their proper sequence, and duration estimating determines the approximate amount of time it would take to perform each activity. At this point, the CPM can be used to determine when each activity can actually be performed.

In any project, there is one sequence of activities that determines the overall duration of the project, known as the **critical path**. A delay in starting any of the activities on the critical path will extend the project duration; conversely, all activities not on the critical path can be delayed for some set period of time without affecting the overall duration. The purpose of the CPM is to determine:

- Which tasks are on the critical path
- The duration of the critical path

- Which tasks are not on the critical path
- How long noncritical tasks can be delayed without affecting the overall project duration

In other words, the CPM tells us, first, which activities are critical and when they *must* take place and, second, the limits for noncritical activities and when they *can* take place.

7.11.1.1 Critical Path Terminology

There are a number of terms associated with the critical path method, which are defined below:

- *Early start (ES):* The earliest an activity can start, based upon the network logic.
- *Early finish (EF):* The earliest an activity can finish, calculated by adding the activity duration to its early start date.
- *Late finish (LF):* The latest an activity can finish without delaying project completion, based on the network logic.
- *Late start (LS):* The latest an activity can start, calculated by subtracting the activity duration from its late finish date.
- *Total float (TF):* The length of time that tasks not on the critical path can be delayed without affecting overall project duration. Some project managers refer to total float as slack time.

    The total float can be calculated by subtracting either the early start from the late start or the early finish from the late finish. Note that activities on the critical path have a total float of zero.

7.11.1.2 Critical Path Notation

There is a standard notation for performing critical path calculations by hand in terms of the manner in which critical path terminology is presented within the precedent diagram boxes. This standard notation is presented in Figure 7.8. Many project management software packages have adopted a similar notation when the logic view screen is selected.

7.11.1.3 Critical Path Calculations

The critical path method embodies two basic sets of calculations: *forward pass calculations* and *backward pass calculations*, as defined below. (Equations will not be provided, as these methods are best explained by way of example.)

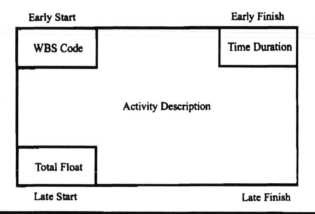

FIGURE 7.8 Critical path activity notation.

Forward pass calculations. Forward pass calculations are carried out first (before backward pass calculations), to determine the early start and early finish dates for each activity. These calculations are based on the assumption that each activity is conducted as early as possible, in accordance with all network logic, including multiple dependencies. The calculations are performed by moving through the network from left to right (start to finish); hence the name forward pass.

Backward pass calculations. Backward pass calculations are carried out to determine the late finish and late start dates for each project activity. These calculations are based on the assumption that each activity is carried out as late as possible, in accordance with the established network logic. These calculations begin with the project end activity and then move through the network from right to left; hence the name backward pass. In order to begin this process, the early finish date of the last activity within the project network is set equal to the late finish date of the last activity.

Some project management textbooks provide a series of equations for performing forward and backward pass calculations, which are usually much more difficult to understand than need be. Here, therefore, the process will be demonstrated by way of example, using the network diagram provided in Figure 7.9A and the notation provided in Figure 7.8. This example is somewhat simplified since there are only three possible paths through the network. The actual process is no different with larger networks, except that more computations are required.

In the following example, note that, in accordance with computer convention, the activity start times denote the beginning of the day that corresponds to the beginning of the activity start date. The activity finish times denote the end of the day. To further simplify hand calculations, it is best to assume that all times are end-of-day times. Therefore, an activity start time of *t* means at the end of working day *t*, which is the same as the beginning of working day *t* + 1. Examples of forward and backward pass calculations are provided in Figure 7.9.

7.11.2   Program Evaluation and Review Technique

Another common method that is useful for schedule development is PERT. While the PERT approach is somewhat similar to the CPM in that it uses sequential logic and also calculates a critical path, the primary difference is that it uses a weighted average duration for each of the project activities (and hence for the project as a whole). This important difference may lead to a more realistic schedule estimate than that produced by the CPM.

When utilizing the PERT approach, three different estimates are provided for each activity duration: optimistic (o), most likely (m), and pessimistic (p). A weighted average or expected time duration (*te*) is then calculated for each task using the equation in Figure 7.10

Note that the expected time (*te*) represents the mean of hypothetical probability distribution (beta distribution). In the development of PERT, it was assumed that a plausible (and mathematically convenient) distribution for the time duration (*t*) was the beta distribution (which has a standard deviation of one-sixth of its range).[10] The equation in Figure 7.10 then provides a linear approximation of the mean of the beta distribution.

One drawback to the PERT approach is that activities with a high level of uncertainty can significantly affect and alter the critical path. Individuals using this approach must therefore be aware of the cumulative effect of activity variance on each path through the network. There are other drawbacks to the PERT approach (more than one critical path, complexity, etc.), but they would require an in-depth discussion which is beyond the intended scope of this section.

7.11.3   Resource Leveling

The third method for schedule development, resource leveling, refers to the process of identifying and solving resource allocation problems. This can best be demonstrated by an example.

In Figure 7.11, the network used from the critical path example has been converted to bar chart format, with all activities set to occur at their earliest start date. In addition, a resource histogram, indicating the number of work hours per day assigned to the project engineer, has been placed directly beneath this project

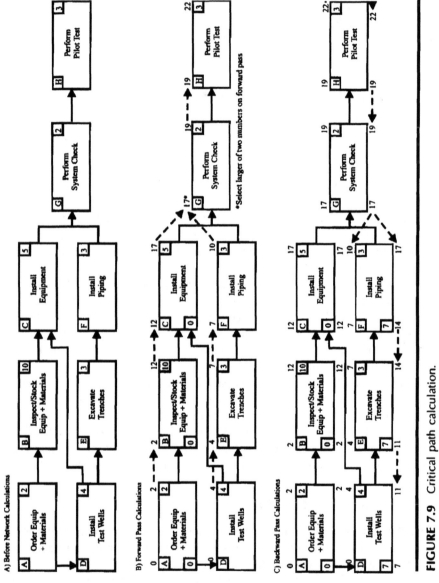

FIGURE 7.9 Critical path calculation.

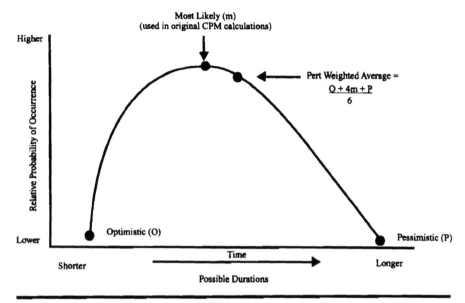

FIGURE 7.10 PERT duration calculation. (From The PMI Standards Committee, *A Guide to the Project Management Body of Knowledge*, PMI Communications, Sylva, NC, 1996, 60. Reprinted with permission of the Project Management Institute, 130 South State Road, Upper Darby, PA 19082, a worldwide organization of advancing the state-of-the-art in project management. Phone: (610) 734-3330, fax: (610) 734-3266.)

schedule. By moving certain tasks within their allotted float, the shape of this histogram can change considerably.

Note in the first histogram, where all activities are set to occur at their earliest start dates, the project engineer is scheduled to work 14 hours on the first day of the project, a clear case of overscheduling. This overscheduling of the project engineer has occurred because task A (order equipment and materials) and task D (install test wells) occur on the same day.

During the planning of this project, it was determined that the project engineer would require two full eight-hour days to make all the arrangements for material and equipment (activity A). It was also determined that on the first day of test point installation, the project engineer would visit the site to confirm test point locations and to perform a quality control check on the installation procedures.

Since the project engineer is overscheduled on the first day of the project, we may want to investigate if we can adjust the activity start dates so as to smooth out the work hours histogram without affecting the overall schedule. From the critical path analysis, it has already been determined that the path consisting of

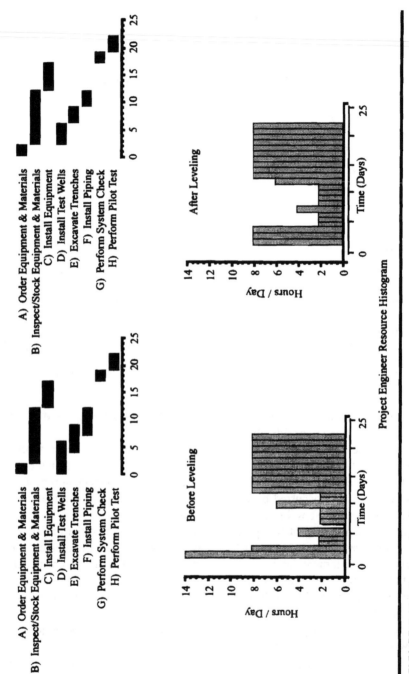

FIGURE 7.11 Resource leveling.

activities D, E, and F contains seven days of total float. Note that by delaying activity D for two days, we can solve the resource scheduling problem without affecting the overall project schedule. (In fact, we have five days of float left within this path, to adjust for other resource allocation problems should the need arise.)

It should be further noted that if the resource allocation problem cannot be solved by using project float, the overall schedule could be affected. In such a case, the project manager would have to choose between a schedule delay and finding an additional resource. On a large job with multiple resources, the resource leveling problem can become quite complex. Adjusting the schedule to level one resource can cause another resource to become overscheduled.

Mathematical techniques have not been discovered that would facilitate programming an optimal solution to this problem. Therefore, attention has been directed to developing heuristic procedures (a series of rules) which will produce "good" workable solutions. These heuristic procedures have names like minimum late start rule and the least slack first rule.

Presented below are examples of heuristic procedures that might be programmed into a project management software package for resource leveling:[11]

1. Give the resource to the activity that has the least float.
2. If two activities have equal float, give the resource to the activity with the longest duration.
3. Where there is conflict between activities, give the resource to the activity that uses the largest amount of the resource.
4. Where there is more than one resource overload in a period, deal with the one that involves the resource of highest priority first.

Each project management package uses its own set of resource-leveling heuristics. Therefore, it is unlikely that each program will reach the same solution. However, the solutions obtained are usually very good and close to optimal.

7.12 EXERCISE

The case study exercise for this chapter involves developing the network diagram and duration estimates for each of the activities identified in the work breakdown structure for the exercises in Chapter 6. Make use of the general process for network diagramming provided in Section 7.9.4 as well as the rules of thumb provided in Sections 7.9.4.2 and 7.10.1. If you have access to a project management software package, such a program will determine the critical path through the network as well as the float associated with the tasks not on the critical path.

If planning is performed in a group setting, the wall planning method can be used to facilitate the process of network diagramming and duration estimating. Work group members should refer to the work breakdown structure as it exists on the wall, to identify appropriate network connectors between the lowest level elements. This process is facilitated if the work breakdown structure is put in the combined work breakdown structure format. In addition, the work groups can record the task duration on each of the cards representing the lowest level elements in the work breakdown structure. Remember that in the standard notation, the durations are placed in the upper right-hand corner of the self-adhesive notes.

Appendix F contains a completed network schedule for the Megacorp former xylene underground storage tank area (i.e., the Best site).

REFERENCES

1. Githens, G.D., Environmental projects in need of better project management, *Project Earth Quarterly*, Vol. 2, No. 3, 1, 1994.
2. Moder, J.J., Network techniques in project management, in *Project Management Handbook*, 2nd ed., Cleland, D.I. and King, R.I., Eds., Van Nostrand Reinhold, New York, 1988, 325.
3. Kelly, J.F., Critical path planning and scheduling: mathematical basis, *Operations Research*, Vol. 2, No. 3, 296, 1961.
4. Lockyer, K.G., *An Introduction to Critical Path Analysis*, 3rd ed., Pitman, London, 1969, 3.
5. Malcom, D.G., Roseboom, J.H., Clarck, C.E., and Fazar, W., Applications of a technique for R and D program evaluation (PERT), *Operations Research*, Vol. 7, No. 5, 646, 1959.
6. Project Management Institute Standards Committee, *A Guide to the Project Management Body of Knowledge (PMBOK)*, *Exposure Draft*, Project Management Institute, Upper Darby, PA, 1994, 64.
7. Project Management Institute Standards Committee, *A Guide to the Project Management Body of Knowledge (PMBOK)*, *Exposure Draft*, Project Management Institute, Upper Darby, PA, 1994, 25.
8. Kerzner, H., *Project Management: A Systems Approach to Planning, Scheduling, and Controlling*, 2nd ed., Van Nostrand Reinhold, New York, 1984, 607.
9. Mulvaney, J., *Analysis Bar Charting: A Simplified Critical Path Analysis Technique*, Management Planning and Control Systems, Washington, D.C., 1989, 12.
10. Moder, J.J., Network techniques in project management, in *Project Management Handbook*, 2nd. ed., Cleland, D.I. and King, R.I., Eds., Van Nostrand Reinhold, New York, 1988, 348.
11. Moder, J.J., Network techniques in project management, in *Project Management Handbook*, 2nd. ed., Cleland, D.I. and King, R.I., Van Nostrand Reinhold, New York, 1988, 345.

Assigning Resources/ Cost Estimating

8

The goal of this chapter is to provide information that will improve cost estimating, thereby increasing the chances of project success and profitability. Poor cost management is one of the most significant causes of project failure, perhaps second only to poor communications management. This underscores the need for strong cost management techniques.

Recall that in Chapter 3 project success was defined as managing all project resources, within the established project constraints — while earning a profit for the performing organization. This definition points out that it is not sufficient to merely develop an accurate cost estimate (i.e., one where the customer's final cost is reasonably close to the original estimate). The cost estimate provided to the customer must ensure that a profit can be earned by the performing contractor.

The development of an appropriate and profitable cost estimate requires the balancing of two opposing forces. On one hand, the planning team wants to develop a higher cost estimate to ensure that there are sufficient funds to achieve the project objectives (including profit). At the same time, the planning team aims to develop a lower cost estimate to compete with other bidding contractors. Arriving at the proper balance between a high versus a low cost estimate can be difficult, yet, the overall success of all service companies and contractors depends on consistently finding this balance.

To facilitate an understanding of cost concepts, this chapter will discuss cost management in general, with special emphasis on assigning resources/cost estimating. (Note that assigning resources and cost estimating are so fully interrelated that they are commonly considered to be one in the same process.)

Once a cost estimate has been approved by operations management, and is ready for submission to the customer, it becomes formally known as the project budget (i.e., step six of the project planning and control cycle [see Figure 3.13]).

This assumes the proposal will be accepted by the customer with no changes to the cost estimate.

The primary topics discussed in this chapter include:

- Project cost management
- Costing versus pricing
- Types of cost estimates
- Information required for assigning resources/cost estimating
- Process for assigning resources/cost estimating
- Guidelines for assigning resources/cost estimating
- Contingency estimating for firm fixed price contracts
- Linear responsibility charts/project plan completion

8.1 PROJECT COST MANAGEMENT

Project cost management is used to ensure that a project is completed within the approved budget.[1] The four processes in project cost management are presented in Figure 8.1.

8.1.1 Resource Planning (Assigning Resources)

The first process in project cost management, resource planning, means determining what resources are needed to complete the project work packages.[1] Resource planning also involves determining what quantities of the various resources are required. The terms resource planning and assigning resources can be used interchangeably.

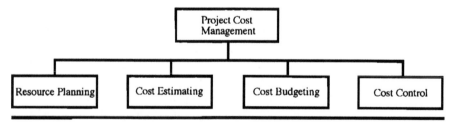

FIGURE 8.1 Project cost management processes. (From The PMI Standards Committee, *A Guide to the Project Management Body of Knowledge,* PMI Communications, Sylva, NC, 1996, 60. Reprinted with permission of the Project Management Institute, 130 South State Road, Upper Darby, PA 19082, a worldwide organization of advancing the state-of-the-art in project management. Phone: (610) 734-3330, fax: (610) 734-3266.)

8.1.2 Cost Estimating

Cost estimating refers to estimating the cost of each of the project work packages.[2] Once the resources have been assigned, cost estimating can be a relatively easy procedure. In order to estimate the cost of each work package, knowledge of the unit cost for each of the assigned resources is required. Unit cost information is available from sources such as company-developed rate sheets (personnel, equipment, materials), laboratory analytical fee schedules, and subcontractor quotes.

8.1.3 Cost Budgeting

Cost budgeting means establishing a cost baseline for each of the project work packages and the project as a whole.[3] Often several iterations of resource planning and cost estimating are required before the final budget is approved. The final budget usually requires approval from senior operations management prior to submittal to the customer.

It is recommended that all budgets have a corresponding time component, as represented by the cumulative cost over time curve (see Figure 3.5). (This curve is often referred to as the project spending plan or "S curve.")

8.1.4 Cost Control

Cost control occurs during project execution rather than planning and is therefore beyond the intended scope of this book. However, Chapter 9, on earned value analysis, provides an introduction to cost control.

Cost control involves monitoring and controlling changes to the approved project budget (baseline cost estimate). Note that a highly detailed proposal/work plan is required for effective cost control. A detailed proposal and work plan facilitate the identification of changes to the scope of work.

8.1.5 Process Overlap

Although the above cost management processes are discussed as separate functions, in practice there is a great deal of overlap. In particular, resource planning (assigning resources) and cost estimating are so tightly linked that they can be viewed as a single process.

The advent of project management software has served to further link the processes of resource planning and cost estimating. This is because these programs allow for the creation of a resource database that includes the unit costs for commonly used resources. Therefore, assigning a resource to a task automatically updates the cost estimate for that task.

8.2 COST VERSUS RETAIL PRICE

Knowledge of the difference between cost and retail price is of great importance. In strict project management terms (as well as traditional business), cost refers to the dollar amount incurred by the seller in providing a product or service. The retail price, on the other hand, is the dollar amount that the seller chooses to charge the customer for providing the product or service. The difference between the retail price and the cost represents the seller's profit or, on an unsuccessful project, the seller's loss.

Although it is not difficult to understand the difference between retail price and cost, it is common in nearly all business endeavors to ask "How much does it cost?" when one is actually referring to *price*. There is no need to try to change this convention, but as a project manager, it is important to know whether one is talking about price or cost. This is because knowing both the cost and the retail price is important to understanding the overall profitability of a project. In some organizations, the retail price is referred to as the external cost estimate and the cost to be incurred by the seller is referred to as the internal cost estimate.

In keeping with conventional terminology, whenever the term *cost estimate* is used, it can be assumed (unless otherwise specified) that reference is being made to an external cost estimate (retail price).

8.2.1 Cost Estimating in an Environmental Service Company

When developing cost estimates in response to a request for proposal, project managers in most environmental service companies are only aware of the retail price. This is because the information available to the project manager, such as personnel rate sheets, standard equipment rental rates, analytical costs, etc., usually reflects only the retail price. Thus, there is no way for project managers to evaluate the actual internal cost of a project or the overall profit potential.

It is important that the project management information system within an environmental service company enable evaluation of the project profit margin based on various retail rate schedules (e.g., standard, strategic, and customer specific). By applying a known, internal, average cost rate for various resources, the system can calculate the potential profit for each project task/resource and the project as a whole.

8.3 TYPES OF COST ESTIMATES

It is common to associate cost estimating only with the proposal process. However, cost estimates can be provided to the customer for reasons other than a proposal (i.e., feasibility studies). There are essentially three types of cost estimates that a project manager can provide to a customer:

- Order-of-magnitude estimates
- Budget estimates
- Definitive estimates

8.3.1 Order-of-Magnitude Estimate

An order-of-magnitude estimate is an approximate estimate, made without the benefit of detailed information. It is usually made during the early stages of an expenditure program, for initial evaluation of a project. Other names for an order-of-magnitude estimate include preliminary, conceptual, factored, and feasibility estimate. The Project Management Institute has defined an order-of-magnitude estimate as one that is accurate within a range of –25 to +75%.[2]

Order-of-magnitude estimates are often produced from cost capacity curves, scale-up or -down factors, and approximate cost capacity ratios. Although items such as cost curves, scale-up factors, etc. are widely available in other industries, such data are not as readily available in the environmental remediation industry.

A common method to develop an order-of-magnitude estimate is through a process known as analogous estimating. Analogous estimating uses the actual cost from a previous or similar project as the basis for estimating the cost of the current project.[4] The accuracy of such an estimate will depend on the similarity between the two projects. Analogous estimating is sometimes called top-down estimating because the estimator may begin with a known dollar amount, based on a similar project, and then allocate it among the various work packages.

Order-of-magnitude estimates are often used in the environmental remediation industry when comparing remediation technology alternatives for feasibility studies. Because these estimates are often prepared for the customer's internal planning, and not for project award, it is common for the customer to pay for such estimates.

8.3.2 Budget Estimates

Budget estimates are often prepared for a customer upon the completion of a remediation system design. These estimates are usually based on flow sheets, layouts, and equipment details. The acceptable accuracy range for a budget estimate is –10 to +25%.[4]

In the environmental remediation industry, customers often request a budget estimate upon completion of the remedial system design. Customers use these estimates internally, to establish funds for the project prior to the release of a request for proposal. It is common for customers to pay for budget estimates.

8.3.3 Definitive Estimate

Definitive estimates are the most detailed cost estimates of all. Their acceptable accuracy range is –5 to +10%.[4] Definitive estimates typically are prepared through the process of bottom-up estimating. Bottom-up estimating involves estimating the cost of the individual work packages and then summarizing, or rolling up, the individual estimates to reach the project total. Definitive estimates are most commonly developed in responding to a request for proposal. Because of the competitive bidding process, customers typically do not pay for definitive cost estimates.

8.3.4 Combining Cost Estimate Types

It is possible, and sometimes necessary, to include both definitive and order-of-magnitude cost estimates in the same proposal submittal. Within the request for proposal in the early stages of a project, customers often request cost estimates for future project stages as well. In effect, the bidding contractors are forced to provide free cost estimates for work that will not be awarded. A contractor that wants to be awarded the current stage of the project must provide cost estimates for later stages. To comply with the customer's request, a definitive cost estimate is supplied for the current stage of the project, and order-of-magnitude estimates are provided for the later stages. Care must be taken to inform the customer that the estimates provided for the latter project stages are order-of-magnitude estimates.

8.4 INFORMATION REQUIRED FOR ASSIGNING RESOURCES/COST ESTIMATING

The type of information required for assigning resources/cost estimating depends on the type of cost estimate one is developing (i.e., order of magnitude, budgetary, or definitive). It is clear that the definitive cost estimate will require the most information of all (especially when compared to the order-of-magnitude estimate).

Accuracy of cost estimating is directly related to project understanding; that is, the greater the project understanding, the greater the chance of accuracy. As discussed previously, project understanding increases with each move along the project planning and control cycle (see Figure 3.13). This is because with each step through the cycle, the planning team has to consider the project in greater depth. Therefore, to ensure an accurate cost estimate, it is recommended that the team have as much project understanding as possible, by formally going through all previous stages of the project planning and control cycle before developing cost estimates.

It is important to realize that when detailed information is gathered early on, the resultant cost estimate will be more accurate. For instance, having the network diagram already prepared will permit easy development of the cumulative cost over time curve, thereby adding a time component to the budget. Additionally, the network diagram and schedule will enable accurate resource leveling.

On some projects, however, it may be possible to develop an accurate cost estimate as soon as the work breakdown structure is in place (especially if the project team has prior experience with the customer and type of project).

Provided below is a list of items that will facilitate an accurate cost estimate. The items of greatest necessity are marked with an asterisk. The remaining items are desirable to improve estimating accuracy, but they are not entirely necessary.

- Project organization chart*
- Work breakdown structure*
- Written scope of work
- Identified specifications*
- Running list of assumptions*
- Duration estimates*
- Network diagram/schedule
- Resource rate sheets (personnel, equipment, labor)*
- Laboratory analytical fee schedules*
- Subcontractor quotes*
- Historical costs for a similar project

Project organization chart. Development of the project organization chart was presented in Chapter 5. The primary reason to have this document on hand is that it will provide a quick reminder of the key individuals to be assigned to the various tasks.

Work breakdown structure. Techniques for developing work breakdown structures were provided in Chapter 6. It should be obvious by this point that the bottom-up approach used to develop a definitive cost estimate cannot be performed without having developed a detailed work breakdown structure.

Written scope of work. The development of the written scope of work was discussed in Section 6.9. In an ideal case, a project manager would have a written scope of work for each work package before attempting to assign resources and estimate costs. During the short time frame allowed for submittal of most proposals, however, it is common for one individual to be writing the scope of work while another is developing the cost estimate. A solution is for both of these individuals to work from the same list of notes pertaining to the scope of work for each work package.

Identified specifications. As explained in Chapter 4, specifications can greatly affect the cost to perform a given task. Therefore, a summary of all project specifications should be on hand whenever cost estimating is being performed.

Running list of assumptions. As discussed in Chapter 6, a running list of assumptions should be documented during the development of the work breakdown structure. These assumptions assist in setting the boundaries for the scope of work and therefore will have an impact on cost.

Network diagram/schedule. Techniques for developing the network diagram and schedule were provided in Chapter 7. As indicated at the beginning of this section, a completed network diagram and schedule should improve the resource-assigning/cost-estimating process. One reason for such an improvement is that additional assumptions (which set the boundaries for the scope of work) are often made during development of the network diagram/schedule. Additionally, the network diagram/schedule helps incorporate time into the project budget.

Duration estimates. Duration estimates, discussed in Chapter 7, figure largely in the cost-estimating process. As explained in Chapter 7, the planning team should develop duration estimates while simultaneously taking resources into account. During the actual assignment of resources, however, these initial duration estimates may need modifying, based on resource availability, resource skill level, etc. (See Section 8.5 for more on duration estimates and assigning resources.)

Resource rate schedules. Resource rate schedules include all company pre-established retail rates for personnel, equipment, and materials.

Laboratory analytical fee schedules. Regardless of the project stage (e.g., extended investigation, pilot testing, installation), there is always a need for sample collection and analysis. Therefore, it is helpful to have on hand the analytical fee schedule for whichever laboratory will be used on the project.

Subcontractor quotes. A quote should be obtained whenever the planning team expects to use a subcontractor for a particular work package. It is never a good idea to assume an amount for a subcontractor without obtaining a quote. This can be a costly mistake.

Historical costs for a similar project. Historical costs from similar projects can be valuable to the cost-estimating process; however, as previously stated, such data are not always available. When using historical costs, be sure to use actual costs for project completion rather than the estimated cost submitted in the historical project proposal.

8.5 IMPACT OF RESOURCE ASSIGNMENTS ON DURATION ESTIMATES

When preparing to assign resources, the duration estimates developed during the scheduling process are vital, yet flexible. As discussed previously, duration estimates are subject to modification based on the actual resources assigned. A simple example is a highly skilled person who is able to complete an activity in less time than another employee, thereby lowering the duration estimate. Another common example is where two people can complete the task in less time than one person alone.

There are two distinct terms related to assigning resources: constant rate assignment and total content assignment. Each has a different bearing on task duration and cost.

A constant rate assignment indicates that the resources will be used on a task for a set number of hours each day (i.e., at a constant rate, which continues for the duration of the task). For example, if it has been decided (through the duration-estimating process) that a task will last for two weeks, assigning an engineer at a constant rate of six hours a day, five days a week will mean the engineer is expected to work a total of 60 hours on that task. Constant rate assignments are typically used on research and development projects.

A total content assignment means that the resource is expected to work a set total number of hours regardless of the duration of the task. For example, assuming that the duration of a task has been estimated as two weeks, a total content assignment of 40 hours for an engineer means that the engineer must finish 40 hours of work within the two-week period.

For the purposes of an environmental remediation service business, assigning resources on the total content basis only is recommended. In addition, it is recommended that the task duration be 20 to 40% longer than the total content duration for the most highly assigned resource. For instance, if it is expected that it will take an engineer 32 hours to complete a task, the engineer should have a total content assignment of 32 hours, while the duration task should be set at 40 hours (a 25% increase). This increase in task duration is a conservative method of creating additional time within a task to allow for any resource delays once the task has been initiated.

8.6 GENERAL PROCESS FOR ASSIGNING RESOURCES/COST ESTIMATING

The following steps represent the general process for assigning resources/cost estimating. It should be noted that, for the most part, these steps can be carried out while working within one of the widely available project management software programs.

1. Gather all necessary information as discussed.
2. Review the work breakdown structure, giving particular attention to the work packages (lowest level elements).
3. Consider each work package and review notes on the scope of work, specifications, and assumptions. Then ask the following questions.
   - What resources are required — in terms of personnel, equipment, materials, and subcontractors — to complete the work package?
   - How many of each resource type will be required to complete the work package and for how long?
4. Assign resources to each work package based on the answers to the above questions.
5. Evaluate the project profit margin if possible.
6. Review the cost estimate for acceptability by considering the guidelines provided in the following section.
7. Compare the cost estimate with similar projects (if data are available).
8. Submit the cost estimate for peer and senior management review, and then adjust as necessary.

8.7 GUIDELINES FOR ASSIGNING RESOURCES/COST ESTIMATING

The following guidelines can assist in the process of assigning resources/cost estimating. These guidelines can also be used when reviewing the completed cost estimate. Note that these are guidelines, not hard and fast rules. Depending on the specific project, it may be necessary (even desirable) to take a different approach.

- The finalized cost estimate should attain the company's minimum acceptable profit.
- Never submit a cost estimate to a customer that involves a financial loss. Many companies will take such an approach with the idea that they will make up for the loss in later project stages. Typically, however, such

companies will not win later project stages, due to their attempt to cut corners, which results in submitting numerous change orders during the project stages they have been awarded. In addition, no one enjoys working on a project that appears to be a financial loser. This causes team morale problems and further financial loss due to reduced productivity.

- Whenever possible, make use of in-house experience. Talk to personnel who may have worked on similar projects in the past.
- Have the people who will be responsible for performing each work package develop the initial cost estimate. Provide such individuals with all necessary information as presented in Section 8.4.
- Make all resource assignments as realistic as possible. Consider the most likely amount of resources needed, with neither an optimistic nor a pessimistic view.
- Do not begin with a final cost in mind. Use the bottom-up estimating approach and then review the final cost for possible adjustments. Any adjustments to the final cost should be the result of a change in the scope of work.
- Assign resources to tasks by the total content approach. Adjust task durations as necessary.
- Resources that are billed at an hourly rate (personnel and equipment) should be assigned to a work package in units of total hours (rather than weeks, months, etc.).
- Use company stock and rental forms as a means of identifying resources that may have been overlooked (e.g., photoionization detectors, interface probes, health and safety equipment).
- Depending on project complexity, figure that project management costs should range between 5 to 15% of the total project cost. Note that due to the large amount of uncertainty, environmental remediation projects seem to require more project management time than typical construction projects.
- For tasks that involve field work, adjust costs for expected changes in levels of personal protective equipment (PPE). See Section 8.8 for more about PPE and costs. Also, be sure to include time and expenses for the associated travel.
- On lump sum contracts, be sure to include a contingency amount. Use one of the approaches for estimating contingency discussed in Section 8.9.
- Be careful to identify who is responsible for the transportation and disposal of hazardous waste from the site, including (but not limited to) soil, groundwater, decontamination liquid, and PPE. If the environmental service company is responsible for making these arrangements, cost quotes from proposed subcontractors should be obtained.

- Maintain a list of assumptions made during cost estimating. This list should be included (along with all other assumptions) in the proposal, as a way of placing boundaries on the project scope of work and limiting overall risk.
- Many of the duration-estimating guidelines presented in Chapter 7 also apply to assigning resources/cost estimating; be sure to review these guidelines as well.

8.8 IMPACT OF PERSONAL PROTECTIVE EQUIPMENT ON COST ESTIMATING

Increased levels of PPE will increase costs due to loss of productivity, additional materials, and additional support personnel.

Loss of productivity. Increased PPE will increase the time for a person to complete a task due to loss of productivity. This increase in time can be represented by productivity adjustment modifiers. The recommended modifiers for Level B, Level C, and Modified Level D are 5.0, 2.5, and 1.25, respectively.

Additional materials. Increased PPE will cause an increase in health and safety materials costs. The unit cost of these materials, per person per day, can be developed based on quotes from material suppliers.

Support personnel. Elevated levels of PPE usually require additional support personnel for decontamination and health-monitoring purposes. Be sure to consult with the project health and safety manager on current policy regarding support personnel for each protective level.

8.9 CONTINGENCY ESTIMATING ON LUMP SUM CONTRACTS

Contingency estimates are usually associated with lump sum contracts. They refer to the dollar amount added to a cost estimate for the purpose of absorbing project risks. The contingency amount is typically not revealed to the customer. In the event that the contract can be executed with little risk impact, the contingency amount is taken as increased profit.

There are a number of methods of estimating contingency amounts for lump sum contracts, ranging from the simplistic to the highly sophisticated. The three primary methods are

- Percentage estimate
- Risk management approach
- Computer modeling

Percentage estimate. The percentage estimate is the easiest and most common way of estimating project contingency. To employ this method, the project planning team will first develop a definitive estimate using bottom-up estimating. The work packages will then be rolled up to arrive at the project cost. The project cost will then be increased by some percentage amount (e.g., 20%). It should be noted that this percentage amount is over and above project profit.

Risk management approach. The risk management approach involves using the procedure discussed in Section 4.12.2. In this case, the planning team identifies all the various risk factors (e.g., risk events, risk probabilities, amount at stake) and then calculates a risk event status for each. Recall that the risk event status equals the risk probability times the amount at stake. The total contingency estimate is derived by adding all the risk event statuses.

This risk management approach can be performed on a work package basis or on the project as a whole. Using this approach on each work package will result in a higher contingency estimate (more conservative) than using it on the project as a whole. In general, the risk management approach results in a higher contingency amount than the percentage estimate approach.

Computer modeling. Computer modeling involves quantitative decision techniques, such as Monte Carlo simulation, decision tree analysis, and range estimating, to arrive at a contingency estimate. These techniques are part of advanced project management and are discussed in Chapter 10. As with any modeling approach, these techniques depend on the skill of the person designing the model. In recent years, the computer-modeling approach has become common in large construction companies.

8.10 PROJECT WORK PLAN COMPLETION/LINEAR RESPONSIBILITY CHART

Once all the resources have been assigned and the cost estimate has been approved, thereby establishing the budget, project planning is complete. Figure 8.2 is included to show how the products of the various steps in the project planning and control cycle (see Figure 3.13) interact to form the complete project plan.

Note in Figure 8.2 how the organizational structure (including the project organization chart) and the work breakdown structure come together to form the

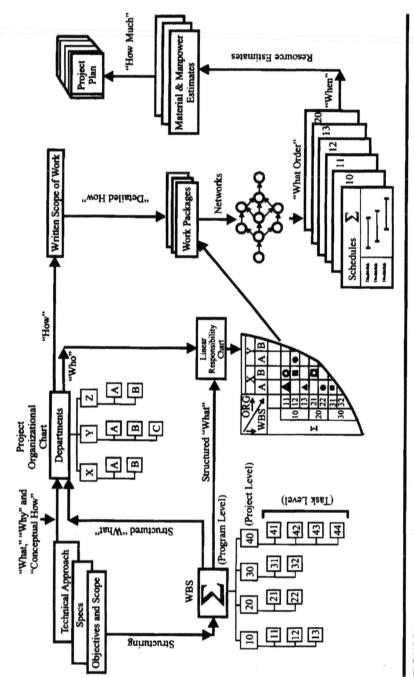

FIGURE 8.2 Integration of project planning elements. (From Kerzner, H., *Project Management: A Systems Approach to Planning, Scheduling and Controlling*, 3rd ed., Van Nostrand Reinhold, New York, 1985, 585. Reprinted by permission of John Wiley & Sons, Inc.)

linear responsibility chart. In addition, the linear responsibility chart feeds into the project work packages.

The linear responsibility chart was first introduced in Section 5.4.2. Once all resources have been assigned to the work packages, it is possible to create an expanded form of the linear responsibility chart. The expanded linear responsibility chart would indicate, for instance, the responsibilities of all project personnel rather than just key personnel. Such a chart allows each of the project participants to see their own responsibilities as well as the responsibilities of others. Under ideal circumstances, all projects would have such an expanded linear responsibility chart.

Another method of communicating responsibilities to the project team members is through the use of work request forms, which in some companies are referred to as work orders or turnaround documents.

Work request forms should include information such as:

- Project and task numbers
- Person assigned
- Planned start and finish dates
- Work instructions
- Equipment required
- Budgeted hours

These work orders can easily be generated once all the steps in the project planning and control cycle have been completed and input into the project management information system. A sample work request form is provided in Figure 8.3.

8.11   EXERCISE

The exercise for this chapter involves assigning resources and developing a cost estimate for each of the work packages identified by the work breakdown structure from the exercises in Chapter 6. In completing this exercise, you may wish to use in-house rate sheets, equipment rental sheets, and material supply sheets. If you do not work for an environmental service company, it is acceptable to assume labor, equipment, and material costs as necessary.

The combined work breakdown structure and time-scaled network diagram produced with the aid of Microsoft Project™ and provided in Appendix G contains costs for the various work packages, which sum to form the various task costs and ultimately the project cost. The costs provided in this example were developed for demonstration purposes only.

Project #   Project Name:

Task:

Sub-subtask:
  Early Start Date: __/__/__   Early Finish Date: __/__/__
  Required Start Date: __/__/__   Required Finish Date: __/__/__

Instruction: _____

Equipment: _____

Resource Category: _____

Budgeted Amount: _____   Scheduled: _____

Project Manager: _____   Date: __/__/__

    Assigned Resource Name: _____

Resource Manager: _____   Date: __/__/__

    Actual Performed: _____

    Estimate to Complete: _____

Signature: _____   Date: __/__/__

FIGURE 8.3 Work request form.

REFERENCES

1. Project Management Institute Standards Committee, *A Guide to the Project Management Body of Knowledge (PMBOK), Exposure Draft*, Project Management Institute, Upper Darby, PA, 1994, 30.
2. Project Management Institute Standards Committee, *A Guide to the Project Management Body of Knowledge (PMBOK), Exposure Draft*, Project Management Institute, Upper Darby, PA, 1994, 31.
3. Project Management Institute Standards Committee, *A Guide to the Project Management Body of Knowledge (PMBOK), Exposure Draft*, Project Management Institute, Upper Darby, PA, 1994, 3.

Earned Value Analysis

9

One of the primary benefits of quality planning is the ability to accurately evaluate progress during project execution. Earned value analysis is used for such evaluations. Throughout Chapters 4 through 8, reference was made to the project planning and control cycle (Figure 3.13). Earned value analysis is the seventh step in the project planning and control cycle.

Earned value analysis techniques are used to identify schedule and cost variances which may occur during project execution. Such variances are usually documented in status reports to operations management and the customer, forming the basis for management decision making (and possibly a new pass through the project planning and control cycle).

The primary topics discussed in this chapter are

- The concept of earned value
- Definition of budgeted cost of work scheduled, actual cost of work performed, and budgeted cost of work performed
- Variance calculations
- Cost and schedule forecasting
- Causes of project variances
- Traditional project planning summary

9.1 THE CONCEPT OF EARNED VALUE

Earned value analysis is a thorough method of measuring project progress. It helps the project manager avoid the common mistake of comparing actual cost expenditures with budgeted expenditures to determine project status. Such comparisons, which are often performed with a simple spreadsheet, can lull both the project manager and the customer into a false sense of security. This is because comparing actual expenditures with budgeted expenditures does not provide adequate tracking information regarding the schedule or value of completed work. (For example, a project may be so far behind schedule that it appears to be under budget.)

Clearly, to obtain a true picture of project progress, a project manager must consider, in addition to cost expenditure, the schedule and the value of the work performed. The concept of earned value analysis considers cost, schedule, and value, in order to obtain a comprehensive picture of project progress.

9.1.1 Schedule Status

It is important to consider schedule status, even on projects where the customer may not be overly concerned with schedule. This is because there is a natural link between cost expenditure and schedule. A project that is behind schedule with *low* cost expenditures is entirely different from one that is behind schedule with *high* cost expenditures. Therefore, development of a planned schedule for cost expenditure (dollars per unit of time), whether in tabular or graphical format (cumulative cost over time curve), is essential for analyzing project progress.

9.1.2 Value of the Work Completed

To effectively control costs, actual expenditures must be compared to some measure of the *value* of the completed work. Considering the value of the completed work brings an important third dimension to the process of determining the project status.[1] Comparing the value of the completed work to the actual expenditures answers an important question: *"What did we get for the money we spent?"*

9.2 DEFINITION OF TERMS

Three terms are central to the process of earned value analysis:

- ■ *Budgeted cost of work scheduled (BCWS)*: Represents the estimated (budgeted) cost of work scheduled to be completed within a given time period.
- ■ *Actual cost of work performed (ACWP)*: Pertains to the actual dollar amount expended on the work completed in a given time period.

■ *Budgeted cost of work performed (BCWP)*: The parameter which deals with the *value* of the completed work; it is the amount of money that was budgeted for the work that was actually completed within a given time period.

It should be noted that each of these values can be determined for any of the work breakdown structure elements, such as tasks, subtasks, subsubtasks, or work packages. As with detailed cost estimating, these values are actually calculated at the work package level and then summed or "rolled up" to provide values for the higher level elements. Also note that all three of these values can be presented in graphical or tabular format.

9.3 UNDERSTANDING BCWS, ACWP, AND BCWP

To further aid in understanding BCWS, ACWP, and BCWP, each of these terms is discussed in detail below.

Budgeted cost of work scheduled. BCWS represents the expected schedule for cost expenditures. A graphical representation of BCWS is provided by the cumulative cost over time curve (see Section 3.8.2.2). Recall that the cumulative cost over time curve was also referred to as the "S" curve because of its characteristic shape.

To determine the BCWS, the project manager must develop a network diagram/schedule for the project in addition to a cost estimate. This is because the network diagram/schedule will provide the necessary information for identifying how much money will be spent on the project within set time periods.

Actual cost of work performed. ACWP is perhaps the easiest value to understand. Collecting ACWP data is easy as well because it is simply the actual amount that has been spent in performing each of the project work packages.

When discussing ACWP, it is important to address the subject of committed costs. Costs are said to be committed when an order is placed (such as purchasing a piece of equipment) or when work is completed (such as drilling a well). Committed costs must be included in the ACWP as soon as they occur, not when an invoice is received from a subcontractor. If such costs are not included in the ACWP at the time they are committed, the variance calculations may indicate that the project is running under budget even if this is not the case. Subsequently, when invoices for the committed costs arrive from subcontractors, the project manager may discover that the project is way over budget.

Budgeted cost of work performed. When individuals are first introduced to BCWP, it is often a difficult concept to understand. However, the concept is not as complex as it might initially appear. Simply stated, the BCWP is what the project manager expected to pay for the work that was actually done.

For example, suppose a project manager intends to install ten monitoring wells at a cost of $5000 per well, representing a total of $50,000 for completing this task. When six of the wells have been installed, the BCWP would be reported at $30,000. This amount would be reported as the BCWP — regardless of the amount that was actually spent in installing the six wells. In this example, it could be said that the six wells have an earned value of $30,000. In fact, some project management texts refer to the BCWP as earned value. Here, however, we will refrain from referring to the BCWP as earned value, to avoid confusing it with earned value analysis, which involves variance calculations and forecasting.

The equation typically used to determine the BCWP is as follows:

$$\text{BCWP} = \% \text{ complete} \times \text{Original budget} \qquad (9.1)$$

While the above equation is suitable for any single work package, the BCWP for the project as a whole is determined by summing or rolling up all individual BCWPs. Note that the dollar value of BCWP as defined by this equation can never exceed the original budget. Also recognize that determining the BCWP with this equation requires a subjective estimate of the level of completion of the work package. Therefore, the accuracy of the BCWP for each work package and the project as a whole depends on an accurate estimate of percent complete. Finally, the original budget is sometimes referred to as the budget at completion.

The difficulty in developing an accurate estimate of percent complete depends on the work package type. In the example above, it is quite easy to determine percent complete (i.e., when six of ten wells are installed, the work package is 60% complete). However, determining the percent complete for items such as a written report, conceptual design, or task management can be quite difficult.

The ability to accurately estimate percent complete is improved by having developed an appropriate work breakdown structure during project planning. This is because proper work breakdown structure design will maximize the ability to start and complete work packages within typical project status reporting periods. Obviously, there is no subjectivity in estimating the percent complete for a work package that is finished. From Equation 7.1, the BCWP for all completed work packages equals the original budget for the work package. The work breakdown structure rules of thumb provided in Chapter 4 will assist in designing an appropriate work breakdown structure.

9.4   VARIANCE CALCULATIONS

Earned value analysis is used to identify schedule and cost variances that may occur during project execution. Having defined the terms BCWS, ACWP, and BCWP, variances can be determined using the following equations.[2]

- *Cost variance (CV) calculation*:

$$CV = BCWP - ACWP \qquad (9.2)$$

   A negative value of CV indicates a cost overrun.

- *Schedule variance (SV) calculation*:

$$SV = BCWP - BCWS \qquad (9.3)$$

   A negative value of SV indicates a behind-schedule condition.

Note that both the schedule and cost variances are defined in terms of cost and make use of the BCWP. A closer investigation of these equations is warranted. Consider first the equation for cost variance (Equation 9.2). This equation provides a true statement of the cost variance since it compares the *value* of the work to the actual *cost* of the work.

The equation for schedule variance (Equation 9.3) is a little more perplexing. One might ask: *What is useful about evaluating schedule variance in terms of cost?* In truth, the equation for schedule variance is not all that useful until it is amended to report the schedule variance as a percentage. Once the schedule variance is known in terms of a percentage (positive or negative decimal value), this percentage can be multiplied by the time since project inception to determine the schedule variance in any unit of time (days, weeks, months). The equations for schedule variance percent and schedule variance time are as follows:[2]

- *Schedule variance percent (SVP)*:

$$SVP = \frac{SV}{BCWS} = \frac{BCWP - BCWS}{BCWS} \qquad (9.4)$$

- *Schedule variance time (SVT)*:

$$SVT = SVP \times \text{Actual duration} \qquad (9.5)$$

Note in Equation 9.5 that by multiplying SVP by the actual duration, the project manager is provided with the truest estimate of schedule status. This is because the equation for SVP (Equation 9.4) makes use of *earned value* (BCWP).

Through the use of BCWP, Equation 7.5 can compare the value of the work completed at a certain point in time to the schedule for cost expenditure at that same point in time.

It is also possible to define cost variance in terms of percent, although it really is not entirely necessary. The equation for cost variance as a percent is as follows.[2]

■ *Cost variance percent (CVP):*

$$CVP = \frac{CV}{BCWP} = \frac{BCWP - ACWP}{BCWP} \qquad (9.6)$$

9.4.1 Performance Indices

Two other values are often computed for schedule and cost variance: the cost performance index and the schedule performance index. The equations for these values are provided below.[2]

■ *Cost performance index (CPI):*

$$CPI = \frac{BCWP}{ACWP} \qquad (9.7)$$

■ *Schedule performance index (SPI):*

$$SPI = \frac{BCWP}{BCWS} \qquad (9.8)$$

A value greater than or equal to one for both of these indices indicates acceptable performance. A value less than one indicates a performance problem.

9.4.2 Sample Variance Calculations

The example in this section is provided to demonstrate the use of the variance equations. The project in this example has a total budget of $100,000 and an expected duration of 12 months. After six months of project execution, BCWS = $58,000, ACWP = $48,000, and BCWP = $32,000.

The earned value analysis is performed using the following equations.

■ *Cost variance (CV):*

$$CV = \$32,000 - \$48,000 = -\$16,000$$

- **Schedule variance percent (SVP):**

$$SVP = \frac{\$32,000 - \$58,000}{\$58,000} = -0.45$$

- **Schedule variance time (SVT):**

$$SVT = -0.45 \times 6 \text{ months} = -2.7 \text{ months}$$

From these equations, it can be seen that the project is $16,000 over budget and 2.7 months behind schedule. Note that if data in this example were analyzed by a standard spreadsheet comparison of actual expenditures versus budgeted expenditures, the project manager could be fooled into thinking that the project is $10,000 under budget.

Calculation of the cost and schedule indices yields the following:

- **Cost performance index (CPI):**

$$CPI = \frac{\$32,000}{\$48,000} = 0.67$$

- **Schedule performance index (SPI):**

$$SPI = \frac{\$32,000}{\$58,000} = 0.55$$

The cost performance index indicates that for every dollar actually spent on the project, only 67 cents in earned value is realized. The schedule performance index indicates that for every dollar budgeted to be spent toward schedule completion, only 55 cents in earned value toward schedule completion is realized.

9.5   GRAPHICAL EARNED VALUE ANALYSIS

Graphical earned value analysis has its basis in the equations provided in the previous sections. Many project managers prefer the graphical approach because it provides a visual picture of the project cost and schedule status which is easy to interpret. In addition, such graphs serve as a method of communicating cost and schedule status to upper level management and the customer. Figure 9.1 presents graphical earned value analysis for the example provided in Section 9.4.2.

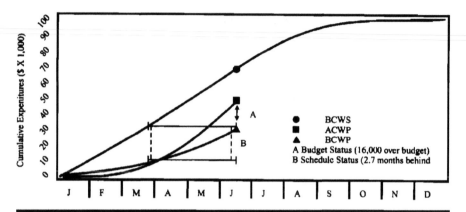

FIGURE 9.1 Graphical earned value analysis. (From Corbitt, R.A., *Standard Handbook of Environmental Engineering,* McGraw-Hill, New York, 1990. 1.37. Reproduced with permission of The McGraw-Hill Companies.)

The cumulative cost over time curve ("S" curve) as shown in Figure 9.1 extends from zero at project start to $100,000 at project completion. With this graphing approach, the cost variance is indicated by the vertical distance between the curves for BCWP and ACWP (distance A), which shows a cost overrun of $16,000. The schedule variance is indicated by the horizontal distance between BCWP and BCWS (distance B), which shows that the project is 2.7 months behind schedule.

9.6 COST AND SCHEDULE FORECASTING

Once a project manager has identified cost and schedule variances, the next logical question is: *What impact will these variances have on the overall cost and duration*? Essentially, there is a need to forecast final project cost and schedule. Often, these forecasts provide the basis for deciding whether or not corrective action is required.

Traditionally in project management literature, the term estimate at completion (EAC) has been used in reference to cost forecasts. A standard term has not evolved which refers to the forecasted schedule for completion. For our purposes, the term estimated duration at completion (EDAC) will be used in reference to the schedule forecast.

It should be noted that it is possible to calculate EAC and EDAC at any level of the work breakdown structure. In practice, these values are determined at the lowest levels of the work breakdown structure and then summed or "rolled up" to determine the higher level values. When reporting EAC and EDAC to the customer (or to upper level management), it is usually best to do so at the project level of the work breakdown structure. This is because positive and negative lower level variances (subtask, work package) can often balance each other out (i.e., cost savings may occur on some tasks to balance out overruns on other tasks). EAC and EDAC reported at the project level provide the best information concerning the overall status of a project. In order to determine the primary causes of an elevated EAC or an extended EDAC at the project level, one can begin by reviewing these values at the lower levels of the work breakdown structure.

9.6.1 Calculation of Estimate at Completion

A number of methods can be used to calculate the EAC, ranging from the simplistic to the highly sophisticated. Each method has its own set of assumptions, advantages, and disadvantages. It should be noted that the more sophisticated methods are not necessarily more accurate. Many project management software packages provide the user with a choice in how the EAC is calculated.

Reviewing all possible methods for calculating the EAC, as well as their associated advantages and disadvantages, is beyond the scope of this book. Three of the most common methods for calculating EAC are presented below and discussed in the following paragraphs.

- *EAC Equation 1:*

$$EAC = ACWP + ETC \qquad (9.9)$$

  where ETC is the project manager's estimated cost to complete the given work package.

- *EAC Equation 2:*

$$EAC = \text{Original budget} - CV \qquad (9.10)$$

- *EAC Equation 3:*

$$EAC = \frac{ACWP}{BCWP} \times \text{Original budget} \qquad (9.11)$$

EAC Equation 1. EAC Equation 1 requires an estimate of the cost to complete (ETC) the work package. This EAC equation is potentially the most accurate, depending on the effort the project manager puts into estimating the cost of the remaining work.[3]

Note that because this equation does not make use of the BCWP, there is no need to estimate percent complete. Many project managers prefer to use EAC Equation 1 based on a belief that it is easier to estimate dollars required to complete a task than to estimate the task percent complete. In reality, ETC and percent complete are closely related, and knowledge of one of the values can be used to calculate the other.

When using EAC Equation 1, a project manager may be tempted to develop one estimate of ETC for the entire project and then add it to the cumulative ACWP for the project at the time of analysis. Although this approach would be easier than (1) estimating an ETC for each work package, (2) calculating an EAC for each work package based on its ACWP, and (3) summing upward to determine the EAC for the entire project, it is also likely to be significantly less accurate.

EAC Equation 2. EAC Equation 2 assumes that current cost variances are atypical and that similar variance will not occur in the future. In other words, this equation assumes that:[4]

- Some cost overruns are unlikely to be repeated.
- Cost overruns that are repeated will be reduced by using experience gained to date.
- Some cost savings will be realized to balance further overruns.

In most cases, these are valid assumptions and the equation can yield accurate results. However, this equation may be a little too optimistic by underestimating the final cost.

EAC Equation 3. This equation makes use of the ratio ACWP to BCWP (the inverse of cost performance index) to develop a multiplier for adjusting the original budget. Note that if the BCWP is less than the ACWP, the ratio will be greater than one, leading to an EAC greater than the original budget.

This equation has been known to overestimate EAC. This is due to multiplication of the original budget by ACWP/BCWP. Essentially, this equation assumes that current variances are typical of future variances. In other words, this equation assumes the reverse of the assumptions associated with EAC Equation 2.

9.6.2 Graphical Determination of Estimate at Completion

It is possible to estimate EAC using the graphical technique by extending the line graphs for ACWP and BCWP and maintaining the variance that occurs between the two at the time of analysis. Using this approach, the EAC would be the end point of the ACWP curve (dollar value). The dollar value at the end of the ACWP curve would be located some distance above or below the end point of the BCWP curve, depending upon the projected variance. In theory, this approach should provide the same result as EAC Equation 2. Note that, by definition, the end point of the BCWP can never exceed the original budget. Figure 9.2 illustrates these concepts.

Two problems are often cited with this approach. The first is deciding what slope to use for projecting ACWP and BCWP. The second problem concerns whether or not it is accurate to maintain a constant variance throughout the projection (it is possible that the variance can increase or decrease over time). However, it can be said that these problems are also inherent in the equations provided in Section 9.6.1.

9.6.3 Calculation of Estimated Duration at Completion

The EDAC calculation forecasts the total time to complete a work package (or the project as a whole) given the amount of work that has been completed at the time of analysis. To arrive at the actual date a project is expected to be completed, a calendar and knowledge of the start date are required in addition to the EDAC. To achieve an accurate EDAC for an entire project, EDACs for each of the project work packages must be determined. However, the EDACs for the work packages cannot be simply rolled up to determine the total project EDAC. This is because some of the work packages may be on the critical path, while others are not. Determining the EDAC for an entire project would be quite time consuming if it were not for the fact that most automated project management information systems can easily perform this calculation.

Similar to estimating the cost at completion, a number of equations can be utilized to calculate the EDAC. The three equations provided in Section 9.6.1 (Equations 9.9 to 9.11) can be altered to arrive at the EDAC.

- **EDAC Equation 1:**

$$\text{EDAC} = \text{Actual duration} + \text{EDTC} \qquad (9.12)$$

where EDTC is the project manager's estimated duration to complete a given work package.

- **EDAC Equation 2:**

$$\text{EDAC} = \text{Original duration} - \text{SVT} \qquad (9.13)$$

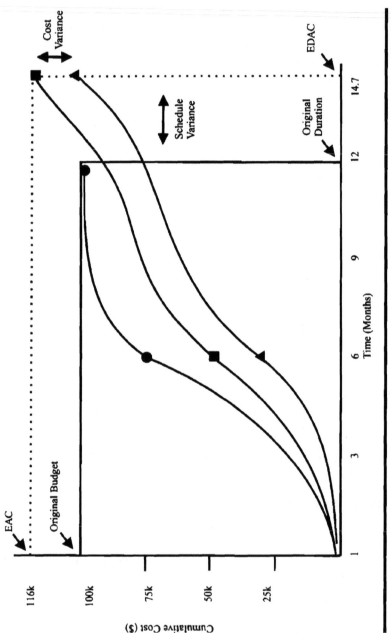

FIGURE 9.2 Graphical project forecasting: EAC and EDAC.

■ *EDAC Equation 3:*

$$EDAC = \frac{BCWS}{BCWP} \times \text{Original duration} \qquad (9.14)$$

It can be seen that EDAC Equations 1, 2, and 3 correspond to EAC Equations 1, 2, and 3, respectively. These corresponding equations have corresponding assumptions. Therefore, the assumptions associated with the EDAC equations will not be discussed here.

9.6.4 Graphical Determination of Estimated Duration at Completion

It is possible to estimate EDAC using the graphical technique discussed in Section 9.5. Using this approach, the EDAC would be determined by the total length of the BCWP curve (time value). The date at the end of the BCWP would be the expected date of project completion. The end point of the BCWP would be located some distance to the right or left of the end point of the original BCWS curve, depending upon the projected time variance. In theory, this approach should provide the same result as EDAC Equation 2.

9.7 CAUSES OF PROJECT VARIANCE

As one might expect, there is potentially a large number of reasons why cost and schedule variances may occur. Mere identification of a variance does not necessarily mean there is a major problem with a project, nor does it mean that immediate corrective action is required.

Determining the need for corrective action requires identifying where the significant variances are occurring, what caused them, and what their impact will be on the final cost and schedule. The question of where significant variances are occurring is answered by reviewing each of the project work packages.

Typical causes of project variance include:

- Failure to understand the customer's requirements
- Inadequate planning, resulting in unnoticed and often uncontrolled increases in the scope of work
- Inadequate work breakdown structure
- Poor cost estimating/unrealistic budget
- Poor duration estimating/unrealistic schedule
- Management problems (skill level and availability of personnel)

- Unforeseen technical problems
- Failure to assess and provide for risks
- Communication breakdown
- Acts of nature

9.8 TRADITIONAL PROJECT PLANNING

This chapter completes the presentation of traditional project planning. A discussion of earned value analysis was included to emphasize the importance of quality project planning. It should be apparent that the earned value analysis techniques will not work well if quality planning was not done in the first place. In other words, in order to determine how well we are accomplishing our plan, we first must have a plan!

Many individuals resist planning. The rationale usually given for such resistance is "things never go as planned." Although it is true that many things do not go as planned, this is no reason for not planning. A well-planned project proceeds more smoothly and succeeds more often than one that is not well planned, even if everything does not go exactly as planned. In addition, project team members typically enjoy themselves more and experience less stress on a well-planned project. Also note that the project management literature is riddled with success stories based on the application of tools and techniques provided in this book.

Although the above benefits are reason enough for creating a sound plan, one of the best reasons may be the ability to identify what went right and what went wrong in terms of the original plan. This information can be used to form a database that will facilitate and improve future project planning. Therefore, it should be understood that the goal is not perfection, but merely continuous improvement.

The tools and techniques provided in this book are processes and guidelines rather than hard and fast rules. It is hoped that many people will be inspired to employ these tools and techniques and even improve upon them.

Recall that project management is about creating change. The project management process provides a method for a group of individuals to pool their talents in a process of creating change to accomplish a goal. Therefore, the tools and techniques provided are applicable to any endeavor, whether business or personal. Individuals have employed these techniques in everything from building a nuclear power plant to staging a theater production and planning a wedding. Those who become skilled in project management will always be able to find a market for their talents, simply because there are so many areas of life in need of change and improvement. It is said that most people fear change. It may be inspiring to realize that modern project management techniques provide tools and techniques for groups of individuals to create change rather than merely react to it (crisis management).

9.9 EXERCISES

Provided below are two exercises that deal with the practical application of earned value analysis. Answers are provided immediately following the written explanation of the exercises.

Exercise 1

You are managing an eight-month project to develop a feasibility study for a large-scale CERCLA site. The budget for the entire project is $300,000. Six months has elapsed since the start of the project. The planned expenditure at the six-month period is $210,000. The total expenditure on the project to date is $225,000. You have met with your task managers, reviewed all work packages, and at this point in time believe that the entire project is 80% complete. Determine the current status (cost and schedule) of the project and forecast the final cost and duration.

Exercise 2

Figure 9.3 shows the BCWS, ACWP, and BCWP curves for four separate projects. Determine the schedule and cost status for each project. Calculations are not required for this exercise; only word statements such as ahead of schedule/over budget will be used.

Answers to Exercises

Exercise 1

$$BAC = \$300,000$$

$$BCWS = \$210,000$$

$$ACWP = \$225,000$$

$$BCWP = \$300,000 \times 0.80 = \$240,000$$

$$\text{Cost variance (CV)} = \$240,000 - \$225,000 = \$15,000$$

$$\text{Schedule variance percent (SVP)} = (\$240,000 - \$210,000)/(\$210,000) = 0.143$$

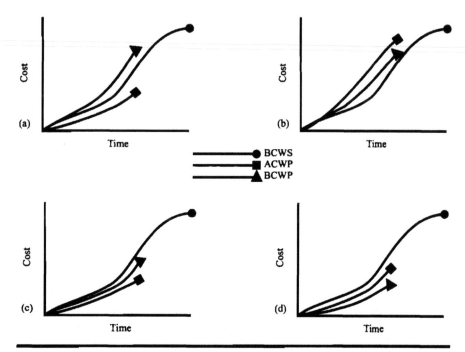

FIGURE 9.3 Graphical earned value analysis: exercise 2. (From Turner, J.R., *The Handbook of Project-Based Management,* McGraw-Hill Book Company Europe, London, 1993, 204. Reproduced with permission of The McGraw-Hill Companies.)

Schedule variance time (SVT) = 0.143 × 6 months = 0.86 months

EAC (Equation 2) = $300,000 – $15,000 = $285,000

EDAC (Equation 2) = 8 months – 0.86 months = 7.14 months

The project is $15,000 under budget and more than three weeks (0.86 months) ahead of schedule. If all continues to go well, the project will finish at a cost of $285,000 and will have a total duration of just over seven months.

Exercise 2

   a. Under budget, ahead of schedule
   b. Over budget, ahead of schedule
   c. Under budget, behind schedule
   d. Over budget, behind schedule

REFERENCES

1. Fleming, Q.W. and Koppelman, J.M., The essence and evolution of earned value, *Cost Engineering*, Vol. 36, No. 11, 21, 1994.
2. Kerzner, H., *Project Management: A Systems Approach to Planning, Scheduling, and Controlling*, 2nd ed., Van Nostrand Reinhold, New York, 1984, 713.
3. Project Management Institute Standards Committee, *A Guide to the Project Management Body of Knowledge (PMBOK), Exposure Draft*, Project Management Institute, Upper Darby, PA, 1994, 33.
4. Turner, R.J., *The Handbook of Project Based Management*, McGraw-Hill, Berkshire, U.K., 1993, 201.

Advanced Project Planning and Risk Management

10

Throughout this book, it has been stressed that environmental remediation projects are fraught with uncertainty. This chapter presents advanced techniques for understanding uncertainty and incorporating it into project plans. The advanced techniques to be discussed make use of a systems modeling approach.

Systems modeling is defined as attempts to develop mathematical models capable of predicting the behavior of real-world activities.[1] A mathematical model that is built around a set of fixed rules such that any given input always results in a specific output is known as a *deterministic model*.[1] If an input can produce a variety of outputs whose selection is determined by rules of probability, the model is known as a probabilistic or stochastic model. This chapter focuses on three systems modeling techniques:

- Decision trees
- Influence diagrams
- Monte Carlo simulation

Decision trees, influence diagrams, and Monte Carlo simulation are all stochastic modeling techniques. In theory, any one of them can be used to solve the same problem, and each will provide probabilistic-based estimates of the cost and time to project completion. In using these techniques, one is attempting to model the situation one is facing by incorporating preferences, options available, and one's beliefs regarding the probability of certain chance events occurring. With the aid of such models, one can evaluate the likely outcome of different strategies and select the strategy that offers the highest probability of providing the desired outcome.

Decision trees and influence diagrams are useful in mapping all decisions and chance events believed to be associated with a given project. In other words, these techniques are useful in defining the structure of the decision problem. This structure can be mathematically analyzed to determine the *optimum decision policy*, which for an environmental remediation project is defined as the best series of decisions leading to the overall lowest cost. The optimum decision policy becomes the basis and the steering mechanism for the overall plan and for developing the work breakdown structure.

In this book, the term *quantitative decision analysis* (QDA) will be used to refer to decision tree analysis and influence diagramming. Some authors also use this term to refer to Monte Carlo simulation, which is acceptable as Monte Carlo simulation can be used to solve decision problems.

A number of software programs have appeared on the market which permit the user to conduct Monte Carlo simulation within a project management software environment. By using these programs, one can conduct a Monte Carlo simulation of a set project plan (in terms of work breakdown structure, network diagram, and assigned resources) to evaluate the expected value of the cost and time to complete a project. In this book, the use of Monte Carlo simulation within a project management software environment will be referred to as *stochastic project modeling* (SPM). (Strictly speaking, however, one could also refer to the influence diagram and decision tree as SPM processes).

This chapter presents a process for combining QDA/SPM as well the traditional project management planning techniques to produce probability-based, often astonishingly accurate, answers regarding the most likely cost and time frame for project completion. A case study example of using this process is provided at the end of the chapter.

Before demonstrating these techniques, this chapter provides information on:

- The importance of the previous chapters
- The history and use of QDA/SPM
- Systems thinking and strategic planning
- The fundamental concepts of probability theory

In order to properly employ the advanced planning techniques of QDA/SPM, it is helpful to understand each of these concepts. As we will see, the mathematics involved in QDA/SPM is not overly difficult. However, application of these techniques requires strong analytical and interpretive skills, as well as significant creativity and intuition.

10.1 IMPORTANCE OF THE PREVIOUS CHAPTERS

The previous chapters provided the foundation for the information presented in this chapter. Chapter 1 provided a summary of the industry trends which are creating the need for early and accurate forecasts of the cost and time to project completion. Chapter 2 investigated the causes of this uncertainty, including items such as heterogeneities in the subsurface environment, limitations of site assessment techniques, and the impact of project stakeholders. In addition, Chapter 2 provided cost-saving guidelines which can be incorporated into the techniques described in this chapter (as well as the planning techniques presented in Chapters 4 through 8).

A more in-depth discussion of uncertainty was provided in Chapter 3. This chapter discussed the fact that a lack of complete information translates directly into uncertainty and further defined uncertainty as the set of all possible chance event outcomes, both favorable and unfavorable. The probability that an outcome will be favorable was referred to as opportunity, while the probability of an unfavorable outcome was identified as risk.[2] *Project risk* was further defined as the cumulative effect of uncertain occurrences adversely affecting project objectives.[2]

In order to apply the advanced project management techniques discussed in this chapter, one must be able to apply the techniques presented in Chapters 4 through 8, including:

- Sorting out and categorizing the vast array of information associated with remediation projects
- Performing stakeholder analysis
- Utilizing the work breakdown structure templates provide in Appendix E
- Developing a network schedule diagram
- Developing conceptual cost estimates

The case study exercises associated with Chapters 4 through 8 were helpful in demonstrating the above techniques. However, a project plan based on these techniques alone is somewhat static, which means that the project plan may not adequately prepare for the impact of chance events that may occur during execution of the project tasks. These chance events can have a significant impact on the overall project scope, budget, and duration. Chance events that could occur in the case study project utilized in Chapters 4 through 8 include:

- The extent of contamination at the site may be greater than originally expected.
- Additional compounds may be found at the site.

- The EPA may issue a consent order for the site before it can be placed in the voluntary cleanup program.
- The pilot study for the high-vacuum dual-phase extraction might fail.

If such chance events occur and the advanced techniques discussed in this chapter are not utilized, a new trip through the project planning and control cycle would be required and perhaps crisis management would kick in. The techniques presented in this chapter account for chance events and their associated costs. In addition, if negative chance events do occur, these techniques permit everyone on the project team to know which corrective actions will be taken. Also, the original project approach can be easily defended.

The fact that chance events will occur in any environmental project, or any other human endeavor for that matter, is often the reason cited for not planning. Many will say, "Why plan when change is inevitable?" The techniques discussed in this chapter, however, represent methods for planning in the face of uncertainty and preparing for project change.

10.2 ORIGINS AND USE OF QUANTITATIVE DECISION ANALYSIS

QDA has been used to solve business planning and public policy problems for over 20 years.[3] The fundamentals of decision analysis are required instruction in top business schools.[3] Although these techniques have been around for some time, they have not been widely utilized other than in the oil and gas exploration industry. There are several reasons for this. The first is that as recently as 20 years ago, computer hardware and software, with the exception of those owned by huge corporations, were not up to the task. Without sufficient hardware and software, a tremendous amount of human effort would be required to perform an adequate analysis. Unless the risks and rewards associated with a particular industry were sufficiently high, there would be no reason to undertake such an effort. The second reason is that managers who received their training at an earlier point in time may not be familiar with these techniques or with interpreting their results; therefore such managers would be reluctant to employ such an approach. The third reason is that in older, more established industries, many companies have developed heuristic methods of managing risks and may be unwilling to try new approaches.

The lack of powerful computer hardware is no longer an issue. The standard desktop computer of the late 1990s is more than adequate to handle even the most complex decision analysis problem. A number of prepackaged computer programs for creating influence diagrams and decision trees became available in the early 1990s. In addition, software packages for performing Monte Carlo simulation with spreadsheets or project management programs have been around since some time in the late 1980s.

Many might wonder why the oil and gas exploration industry was an early proponent of decision analysis techniques. The answer is easily found in the risks and rewards associated with discovering a highly productive oil field. Such a discovery could be worth hundreds of millions of dollars. Because of this, exploration and production companies are constantly in search of promising locations to explore for oil. If a company decides to drill an exploratory (so-called wildcat) well in a given area, the decision is typically based on limited information (which by definition creates uncertainty and risk). The cost of the drilling lease can be on the order of many millions of dollars. Drilling rig mobilization costs and daily rates are also extremely high.

Suppose a company finances the drilling of a 15,000-foot well in the Gulf of Mexico, and after reaching the planned depth, no oil is found. Some decision makers may feel that, given the investment, it would be best to drill another 1000 feet just to make sure nothing was missed (the problem with this is where to stop with such logic). Other decision makers might feel that, in order to prevent additional losses, it is best to stop now. Still others might feel that the best thing to do is gather more information, perhaps by running a suite of downhole geophysical logging tools, before giving up. Who is right? The answer depends on the potential reward if the different choices result in a positive outcome. Hence the need for QDA.

10.3 UNDERSTANDING QUANTITATIVE DECISION ANALYSIS

The techniques of QDA are gaining increased acceptance in the environmental remediation industry, particularly in the areas of brownfield redevelopment and lump sum cost to closure bidding. Decision makers who have adopted these methods have a good understanding of what the techniques of QDA are and are not. These techniques are:

- Disciplines for helping decision makers choose wisely under conditions of uncertainty[4]
- Methods for thinking systematically about complex problems[5]
- Information sources that provide insight into the situation, risks, objectives, and trade-offs[4]

These techniques are not:

- Taskmasters that provide results which must be accepted blindly[4]
- A replacement for intuition[4]
- A competitor to personal styles of decision making[4]

With these points clearly understood, QDA can be seen as methods for augmenting and improving the standard decision-making processes, not replacements for them. When first exposed to these techniques, many people wrongly assume that they limit creativity. However, experienced decision analysts know that developing an appropriate, representative systems model requires a great deal of experience, creativity, and intuition.

10.4 USES OF ADVANCED PROJECT PLANNING TECHNIQUES

Project planning processes that involve QDA/SPM can be utilized to:

- Ensure that the best overall project plan is in place
- Make probability-based estimates of the cost and time to achieve project closure
- Estimate environmental liability for Securities and Exchange Commission reporting
- Establish a basis for cost cap insurance
- Develop accurate environmental budgets for corporate environmental management groups
- Determine the bid price for lump sum cost to closure projects

10.5 SYSTEMS MODELING

As previously stated, the application of QDA/SPM requires a systems engineering or modeling approach. When first introduced to QDA/SPM, some people are skeptical about whether or not it is possible to develop such models. A common criticism is that there are just too many parameters to create an accurate model. However, the same criticisms can be (and have been) applied to models designed to determine the fate and transport of contaminants in the subsurface environment.

Those deeply involved with fate and transport modeling know that their process is a both a science and an art. These models involve identifying the major parameters of a problem and understanding the mathematical relationships among them. As discussed in Chapter 2, major parameters include items such as contaminant vapor pressure, solubility, viscosity, and biodegradability, as well as the nature of the subsurface environment, including porosity and permeability (to name only a few). The relationships among these parameters are the various equations that deal with dispersion and partitioning. Ensuring that a fate and transport model is representative involves experience, intuition, and testing. Therefore such models are often built in phases, from a conceptual (perhaps one-dimensional) to a refined (perhaps three-dimensional) version.

The same approach is applicable to QDA/SPM. The number of parameters involved in modeling a project is neither as difficult to identify nor as large as one might expect at the outset. The project modeler knows at the outset that:

- Certain standard tasks must be performed in completing a project.
- There are choices in the way these tasks are implemented (e.g., soil gas survey, geoprobe).
- The tasks are performed in a certain common sequence.
- Common chance outcomes are likely to occur as a result of performing standard tasks, such as:
  □ Larger extent of contamination than expected
  □ Pilot test results not favorable
  □ Toxicological risk assessment results not approved by regulators
  □ Stakeholder group reacts negatively to proposed remedial action plan

Once chance events have been identified, the modeling team assigns a probability that the chance events will occur, as well as a dollar value if they should occur. Therefore, the process requires estimating both the probability and the cost of chance events. This is another area where individuals may react negatively to the use of the processes discussed in this chapter. The objection is that the assigned probability is subjective and that conceptual cost estimates can be inaccurate.

The easiest response to such objections is that project managers make subjective judgments every day, such as when they assume they will be able to complete a project within a given dollar amount or schedule. However, with the traditional approach, such decision makers may not take the time or devote the effort to evaluate their probability assignment in detail. The advanced project planning approaches are much more rigorous (i.e., a planner is able to clearly see the impact of different assignments of probability and cost). Later in this chapter, we will see that it is possible to perform a sensitivity analysis on both the probabilities and cost assignments in order to:

- Better understand the situation we are facing
- Determine how far off assignments of cost and probability can be and still be assured we are following the right plan
- Improve the overall quality of the model

10.6 SYSTEMS THINKING AND STRATEGIC PLANNING

It should be clear that developing representative project models requires the ability to think in terms of systems. What exactly is a system? If we were to look this word up in a dictionary, we would find a definitions such as a system is a group of elements that function together as a whole. This definition, although adequate, does nothing to inform as to how to perform systems thinking.

Engineers who deal in thermodynamics have a more accurate definition of a system. A thermodynamic system is defined as the matter enclosed within an arbitrary boundary but a precisely defined control volume.[6] Everything external to the system is defined as the surroundings, environment, or universe.[6] The environment and the system are separated by the boundaries.[6] This definition is a little more helpful in that we come to realize that a system has boundaries that separate it from its surroundings. It should be noted that this does not mean that the surroundings cannot act or impinge on the system (recall stakeholder analysis) or that the system cannot act on the surroundings.

A more comprehensive definition of a system is provided by R. Buckminster Fuller in the book *Synergetics*. According to Fuller, "A system is the first subdivision of (the) Universe into a conceivable entity separating all that is non-simultaneously and geometrically outside the system, ergo irrelevant, from all that is geometrically inside and irrelevant to the system; it is the remainder of (the) Universe that conceptually constitutes the system's set of conceptually tunable and geometrically interrelatability of events."[7] Although this definition may seem confusing at first, it says a lot about the systems thinking required to develop a representative project model. With this definition, Fuller states that a system *is the structure* itself which results from identifying the relevant set of events and their relationships. In Fuller's definition, we find discussion of events inside the system as well as those outside of it, similar to the thermodynamic definition of a system. However, Fuller's definition mentions irrelevant events.

For the purposes of project modeling, events can be defined as the various chance occurrences (chance events), choice options (decisions), and outcome values (physical data of monetary amounts). Systems thinking, according to Fuller, is the conscious dismissal of "irrelevancies."[8] These irrelevancies can be placed in two categories: those too large or too infrequent to influence the problem at hand (by definition, outside of the system) and those too small or too frequent to play a part (insignificant and inside the system). The systems thinking Fuller describes is similar to tuning a radio by dismissal of irrelevant (other-frequency) events.[9] This dismissal ultimately leaves a smaller number of events relevant to the problem at hand. The solution to a problem involves identification of the relevant set of events and then determination of their interrelationships. Fuller would often explain that a comprehensive solution to a problem can be envisioned as a geometric polyhedral structure, where the edges represent relationships and the vertices represent events.

To demonstrate this point, Fuller would use a number of concentric circles, presented as Figure 10.1. Outside of the outermost circles are those events that are too large and too infrequent to be relevant. Inside the next circle are those events which are almost or "tantalizingly relevant." These events present a problem because one has to decide whether or not they are relevant. Anyone who has tried to develop a plan in a group setting has no doubt found that certain members of a group may think that a particular issue (i.e., event) is extremely relevant, while others feel it is not significant at all. Moving toward the center of our diagram, we encounter those events which we are certain are relevant. It is these events for which we seek interrelationships (i.e., connect by an influence diagram or a decision tree). Moving further in, we encounter another set of so-called "tantalizingly relevant events," but this time on the micro scale. Finally, in the innermost circle, are the insignificant micro irrelevant events. Fuller would envision each of the relevant events as the vertices of a polyhedral structure, with each of the edges of the structure representing the relationship between

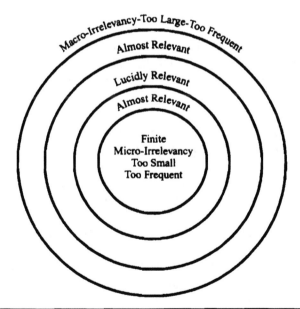

FIGURE 10.1 Systems identification of relevant events. (From Fuller, R.B., *Synergetics — Explorations in the Geometry of Thinking*, Macmillan, New York, 1975, 235. Used with permission from the estate of R. Buckminster Fuller. For more information about Buckminster Fuller's work, visit www.bfi.org.)

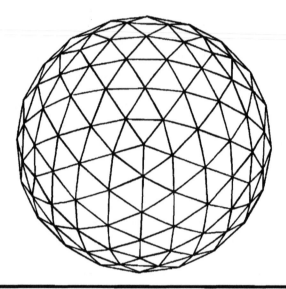

FIGURE 10.2 Polyhedral representation of system structure with vertices representing relevant events and edges representing relationships.

events (see Figure 10.2). In our case, we will use the decision analysis techniques to relate the events and develop the model. By mathematically analyzing this model, we arrive at the strategic plan or the optimum decision policy.

10.7 FUNDAMENTAL CONCEPTS OF PROBABILITY THEORY

The techniques of QDA/SPM draw heavily on probability theory. In fact, one of the central principles of decision analysis is that uncertainty can be represented through the appropriate use of probability.[10] Therefore, before proceeding with a demonstration of QDA/SPM, it is helpful to review some of the basics of probability theory.

In probability theory, the act of conducting an experiment (trial) or taking a measurement is known as sampling.[11] Probability theory determines the likelihood that a particular event will occur.[11] An event (*e*) is one of the possible outcomes of a trial. It is important to note that an event can be numerical or nonnumerical, discrete or continuous, dependent or independent.

An example of a nonnumerical event is the toss of a coin.[11] The roll of a pair of dice is a discrete numerical event, since only certain numbers can result. A large number of cost estimates from a variety of independent sources, perhaps for installing an air-sparging system, is an example of a continuous numerical event, since the estimates can take on any value (within reasonable limits). Independence means that the result of a particular run of an experiment is not affected by previous runs of the experiment. Referring to the air-sparging cost estimates example, we could say that the various results are independent if the individuals providing the estimates were not influenced by or had no knowledge of estimates provided by other parties.

Taken together, all of the possible events in a given experiment constitute a finite sampling space, which can be defined as $E = (e_1, e_2, ..., e_n)$.[11] For example, given a pair of dice, the finite sampling space is E = (2, 3, 4, 5, 6, 7, 8, 9, 10, 11, 12), since these are the only possible numbers that can result from a single trial or roll of the dice. The probability of event e_1 occurring is designated $P(e_1)$ and is calculated as the ratio of the total number of ways that the event can occur to the total number of outcomes in the sample set.[11]

Referring again to rolling a pair of dice, we know that each die has six faces, with the dots on each face representing the numbers one through six. On any given roll (of both dice), one of the six faces will turn up on each die. Therefore there is a total of 36 possible outcomes (6 × 6 = 36). As previously stated, the total number of dots on the two turned-up faces can only add up to any one of the numbers 2 through 12 (finite sampling space). Let's say we wanted to determine the probability of rolling the number seven, $P(7)$. There is a total of six ways to produce the number seven by rolling two dice (see Figure 10.3). Therefore, $P(7)$ = 6 ÷ 36 = 0.167. The probabilities of rolling the numbers 2 through 12 with a pair of dice are presented in Table 10.1.

There are two important points to note about Table 10.1. The first is that all of the probabilities are between zero and one, which is a fundamental require-

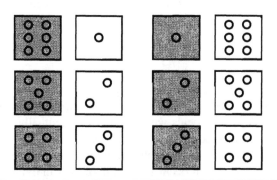

FIGURE 10.3 Ways to produce the number seven.

TABLE 10.1 Probabilities of Rolling the Numbers 2 through 12

$$P(2) = 1/36 = 0.0278$$
$$P(3) = 2/36 = 0.0556$$
$$P(4) = 3/36 = 0.0833$$
$$P(5) = 4/36 = 0.1111$$
$$P(6) = 5/36 = 0.1389$$
$$P(7) = 6/36 = 0.1667$$
$$P(8) = 5/36 = 0.1389$$
$$P(9) = 4/36 = 0.1111$$
$$P(10) = 3/36 = 0.0833$$
$$P(11) = 2/36 = 0.0556$$
$$P(12) = 1/36 = 0.0278$$
$$1.0000$$

ment of probability. This fundamental requirement is defined mathematically in Equation 10.1:

$$0 \le P(e_i) \le 1 \tag{10.1}$$

The second important point is that the sum of the probabilities of rolling the numbers 2 through 12 is one. This is a fundamental requirement for probabilities of a set of events which are *mutually exclusive* and *collectively exhaustive*. Mutually exclusive means that one and only one possible outcome can occur in a given trial. This is obviously true for a single roll of the dice where the sum of the two die can result in only one of the numbers 2 through 12. Collectively exhaustive means that there are no possible outcomes other than those in our set. This is also true for a pair of dice (i.e., it is not possible to roll the number 13). The second fundamental concept of probability theory for mutually exclusive and collectively exhaustive events is mathematically stated in Equation 10.2:

$$\sum_{i=1}^{n} P(e_i) = 1 \tag{10.2}$$

Equation 10.2 is a fundamental rule of probability dealing with a combination of events; in other words, a rule of *joint probability*. It is actually an extension of the next rule, which deals with a subset of n mutually exclusive events from a finite sampling space. This rule, mathematically defined in Equation 10.3,[11] states that if two or more events are mutually exclusive, the probability that any one of the events will occur is the sum of their individual probabilities.

$$P(e_1 \text{ or } e_2 \text{ or } \ldots e_k) = P(e_1) + P(e_2) + \ldots + P(e_k) \tag{10.3}$$

Note that the word *or* is an important term in understanding the equation. If we were rolling a pair of dice and wanted to know the probability of rolling a five, seven, or ten on our next roll, the answer is found by summing the individual probabilities: $(0.111 + 0.167 + 0.083) = 0.361$.

The last fundamental rule also deals with joint probability of two *independent* events either from the same sampling (as long as replacement occurs, such as drawing a card and returning it to the deck) or from different sampling spaces. If the events are from the same sampling space, this law applies to trials taken in series, as in the probability of rolling a three on one roll and a six on the next roll. If the events are from different sampling spaces, E and G, this rule gives the probability that e_1 and g_i will both occur. Note that since the events are from different sampling spaces, there is no reason why they cannot occur simultaneously. This rule is mathematically stated for events from different sampling spaces in Equation 10.4:[11]

$$P(e_i \text{ and } g_1) = P(e_i) \times P(g_i) \qquad (10.4)$$

The discussion of influence diagrams and decision trees later in this chapter will make repeated use of these concepts, in particular Equations 10.1, 10.2, and 10.4. In developing influence diagrams and decision trees, one must identify outcome events associated with a given chance node. The outcome events must be mutually exclusive and collectively exhaustive. Each outcome event, at each chance node, will have an assigned probability; the sum of these assigned probabilities must be one.

10.7.1 Describing Experimental Data

Probability theory is used in describing experimental data (i.e., outcome events). Describing experimental data involves the use of graphing techniques as well as a number of calculated descriptive statistics which measure the central tendency of data and their degree of dispersion. Familiarity with common techniques for describing experimental data is important in understanding the results of a stochastic project model.

10.7.1.1 Graphical Representations of Experimental Data

Common graphical representations for describing experimental data include the frequency histogram, the probability (density) distribution, and the cumulative probability density curve. A frequency histogram is created by first dividing the value range into intervals and then counting the number of occurrences within each interval.[12] Finally, the histogram is created by drawing vertical bars with heights proportional to the number or frequency of occurrences within each interval. Figure 10.4 presents a frequency histogram for an ideal experiment involving 36 roles of a pair of dice. Note that the frequency on the left-hand side is equal to the probability of rolling a particular number and that the sum of all of the rectangular areas is equal to one.

Figure 10.5 is a fictitious histogram which might result from 25 independent estimates of the cost to install a soil vapor extraction/air-sparging system at a given site, where the collective estimates range from $110,000 to $185,000.

If it were possible to collect many independent estimates (perhaps 1000 or more) of the cost to install the soil vapor extraction system and a very small class interval was utilized, the histogram might take on a smooth continuous appearance,[12] as shown in Figure 10.6.

The curve shown in Figure 10.6 is referred to as the *probability density function*. A probability density function, $f(x)$, gives the probability that discrete event x will occur. In other words, for the curve shown in Figure 10.6, we could say that $P(x) = f(x)$. The probability density function can take many different forms. In fact, many theoretical probability distributions have been identified to model certain systems. Figure 10.7 presents some of the more common probability density models. Section 10.10 on Monte Carlo simulation discusses the application of these models.

The continuous distribution function, $F(x)$, gives the probability of a numerical event being less than or equal to x. If the probability density function is known, the continuous distribution curve can be developed by summing the area under the curve from left to right, which is the same as integrating the probability density function. In other words, if $f(x)$ represents the probability density function, the continuous distribution curve can be expressed mathematically, as indicated below:[13]

$$F(x) = P(X \leq x) = \int f(x)dx \qquad (10.5)$$

Figure 10.8 further illustrates the relationship between the probability density function and the cumulative probability distribution curve.

The cumulative probability distribution curve, when demonstrating project modeling results, is also referred to as the risk profile and the cumulative probability versus cost curve. It provides a significant amount of information in one easy-to-read graph.

10.7.1.2 Measures of Central Tendency

The central tendency or "location of the data" refers to a value which is typical of all sample observations. Three common measures of central tendency of data are the mean, median, and mode. The mean is what most people call the "av-

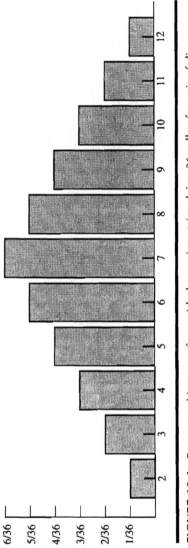

FIGURE 10.4 Frequency histogram for an ideal experiment involving 36 rolls of a pair of dice.

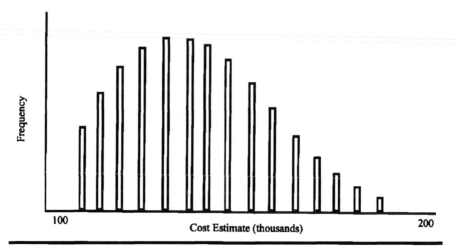

FIGURE 10.5 Frequency histogram of 25 soil vapor extraction/air-sparging cost estimates.

erage." It is the sum of all measured values divided by the number of measurements. A "weighted average" is actually a weighted mean. An important measure of central tendency in project modeling is the *expected value*. The expected value is the probability weighted average of all possible project outcomes. In other words, the expected value is simply the mean. It is the average of all project costs if the project is performed in accordance with a certain strategy.

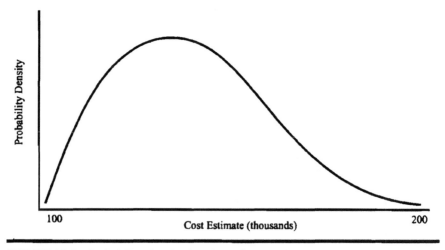

FIGURE 10.6 Probability density function of cost estimates.

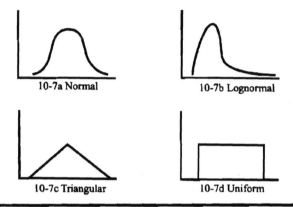

FIGURE 10.7   Examples of common probability density distributions.

The expected value is important in that it is an unbiased estimator. As such, the expected value is the best number to use when comparing alternative strategies.

The mode is the value that occurs more frequently than the other values. The mode is where the concentration of the data is the greatest. The mode can be read directly from a histogram or from a probability density function. Its value is that it is a relatively quick measure of central tendency.

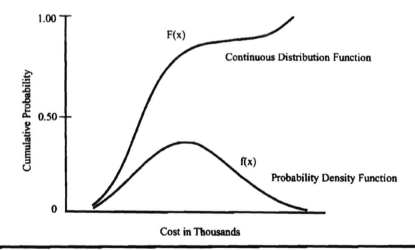

FIGURE 10.8   Relationship between the probability density function and the continuous distribution function. (Reprinted from Michael R. Lindeburg, *Engineer-in-Training Reference Manual,* 8th edition, copyright ©1988 by Professional Publications, Inc., Belmont, CA.)

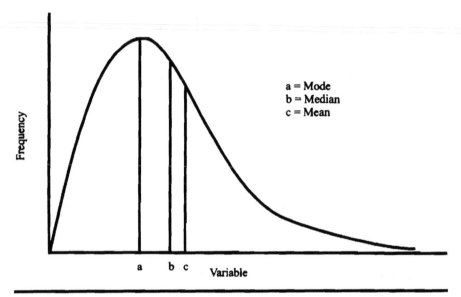

FIGURE 10.9 Location of mean, median, and mode for a log normal distribution.

The median is the point in a frequency histogram (or probability density function) that partitions the total set of measurements into two sets equal in number. The median is the middle point of all observations. The median can be determined by counting from either end of a frequency histogram until half of the values are accounted for. Percentile rank is related to the concept of the median. The median could also be called the 50th percentile rank.[14] In a similar fashion, the 90th percentile rank would be that point for which the cumulative frequency (probability) is 90%.

The mean, median, and mode for a normal (Gaussian) probability distribution are all the same value. However, for most other distributions, they represent different values. Figure 10.9 indicates sample locations of the mean, median, and mode for a log normal probability distribution.

10.7.1.3 Measures of Dispersion

Although the mean, median, and mode provide information about the central tendency or the "location of the data," they indicate very little about the way in which the data are spread out or dispersed. Figure 10.10 illustrates why having an idea of the dispersion of the data is important to project modeling. This figure shows two different projects which have the same mean (expected value) even though their probability distributions are significantly different. Because project

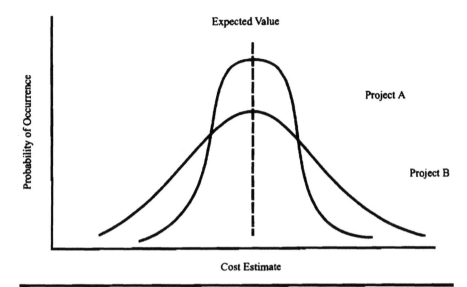

FIGURE 10.10 Probability distribution for two different projects; project B involves more risk. (From Wideman, R.M., *Project and Program Risk Management: A Guide to Managing Project Risk and Opportunities,* Vol. 6, Project Management Institute, Drexel Hill, PA, 1992, V-3. Reprinted with permission of the Project Management Institute, 130 South State Road, Upper Darby, PA 19082, a worldwide organization of advancing the state-of-the-art in project management. Phone: (610) 734-3330, fax: (610) 734-3266.)

B has a wider distribution of data, one could say that it is a riskier project and would require a greater contingency amount than project A to provide the same confidence level for not exceeding the budget.

The three common measures of dispersion are the range, sample variance, and standard deviation. These parameters can be understood as indications of the dispersion of data about the central location, which for practical purposes is the sample mean. The range is the simplest measure of dispersion and is found by subtracting the largest sample value from the smallest sample value. The range can be greatly affected by extreme values (i.e., low probability values) and therefore is somewhat limited.

The sample variance, also known as the mean squared deviation, is more useful than the range because it considers all of the sample values. The equation for sample variance is

$$s^2 = \frac{\sum_{i=1}^{n}(x_i - u)^2}{n - 1} \tag{10.6}$$

In this equation, u represents the mean of the sample set and n represents the total number of samples in the set. The larger the variance, the greater the degree of dispersion. Therefore, the variance is useful when comparing two sample sets. For the project examples in Figure 10.10, the calculated project variance for project B would be greater than that for Project A.

The sample variance is measured in squared units of the sample population. If the samples are measured in inches, the variance would be in inches squared. By the same token, if the samples are measured in dollars, the units of variance would be dollars squared. As one might imagine, it is much more useful to have the measure of dispersion in the same units as the sample set. This is obtained by taking the square root of the variance, which is known as the *standard deviation*. The equation for standard deviation is

$$ s = \sqrt{\frac{\sum_{i=1}^{n} (x_i - u)^2}{n - 1}} \tag{10.7} $$

To further understand standard deviation, an example involving normal (Gaussian) distribution is applicable. Many people are familiar with normal distribution because it represents the outcomes of a variety of phenomena, in particular student test scores. In a normal distribution, one standard deviation from the sample mean will contain 68.3% of all outcomes. Two standard deviations from the sample mean will contain 95.4% of all outcomes, and three standard deviations will contain 99.7% of all outcomes.

10.8 ELEMENTS OF A DECISION PROBLEM

In any decision analysis problem, there are essentially three possible elements (or, as stated earlier, events): decisions, chance events, and outcome values. An important point to bear in mind is that we usually have no control over whether or not a certain chance event will occur. In addition, we have only limited control over the cost to perform corrective action if a certain negative chance event does occur. The only real control we have is over the decisions we make. However, once we make a decision, we open ourselves up to a number of chance events, each with its own outcome values and each creating the need to make new decisions. For example, on an environmental remediation project, we may choose to implement a remedial technology such as bioventing. If this technology allows us to achieve our cleanup goals within our estimated time frame, perhaps no additional decisions will be required. However, the technology may not perform as well as we had hoped. At this point, we have new decisions to make, such as upgrading the process or seeking approval from a regulatory agency for less stringent cleanup goals. This illustrates the sequential nature of decision that we must capture in our project model.

Many situations have as their central issue a decision that must be made immediately. In other words, we are dealing with important decisions that must be made sequentially earlier in time. These important decisions are the ones we are most concerned about and, as we will see both in the next section and in the case study example, are the ones that are at the beginning of or initiate an influence diagram or decision tree.

10.9 STRUCTURING THE DECISION PROBLEM

The two primary ways of structuring a decision problem are the decision tree and influence diagram. Both techniques provide a graphical representation of the problem structure, and each has its own set of advantages and disadvantages in modeling difficult decisions. The advantages and disadvantages of each method are best understood by way of demonstration. We will begin with the decision tree.

10.9.1 Decision Trees

Decision tree analysis involves mapping the various decisions and chance events associated with a project in a diagram that branches and expands, from left to right, in a tree-like manner. In the diagram, squares are used to represent decisions and circles represent chance events. These shapes are referred to as nodes (i.e., decision nodes and chance nodes). Both the squares and the circles can have any number of branches emanating from them.[15] The branches emanating from a square correspond to choices that are available to the decision maker.[15] The branches emanating from a circle represent possible outcomes of a chance event.[15] Costs (or profits) are assigned to each choice branch (event), and both probabilities and costs (or profits) are assigned to each chance branch (event).

When developing a decision tree, the branches emanating from the decision nodes must be such that one and only one option can be chosen. In other words, the decision options must be *independent* and *mutually exclusive*.

In a similar fashion, the branches emanating from a chance node must correspond to a set of *mutually exclusive* and *collectively exhaustive* events. Therefore, in accordance with Equations 10.1, 10.2, and 10.3, the probabilities associated with the branches emanating from each chance node must add up to one.

Each path through the tree represents a specific combination of decision alternatives and chance event outcomes.[16] Such a path represents a single possible state in the world or a scenario.[16] Mathematical analysis of the decision tree is done by first moving through each pathway from left to right and summing up the costs associated with the choice and chance events until a definite end point (outcome) is reached. Thus, the total cost for each pathway end point is determined. The probability of any one end point occurring is determined by multiplying the probabilities along a given pathway moving from left to right (in accordance with Equation 10.4).

Once the pathway end-point costs have been determined, the tree is mathematically analyzed in reverse (from right to left) in a process known as folding back the tree. The expected value of each chance event is determined by multiplying the various branch costs by their associated probabilities and then summing the result. In other words, the expected value for a given chance node is the *probability weighted average* (i.e., the mean) of the branch outcomes. This mean is referred to as the *expected monetary value* (EMV). When choice nodes are encountered, as a result of folding back the tree, the EMV for each branch is compared. Since in the environmental industry we are typically dealing with costs (rather than profits), the branch with the lowest EMVs is chosen. Branches not chosen are eliminated from further consideration, in a process referred to as pruning the tree. The series of decisions leading to the lowest EMV is known as the optimum decision policy. This policy represents the basis of the strategic project plan.

10.9.1.1 Analyzing a Basic Risky Decision with a Decision Tree

A basic risky decision is one that involves a single decision, one uncertain event, and one outcome value (cost) which is affected by both.[17] This statement is clarified by referring to Figure 10.11 and later by Figure 10.17. This figure depicts a decision tree for a basic risky decision in an environmental remediation project. In this example, the decision is between two different soil remediation technologies. Therefore, there are two branches emanating from the remedial technology choice node. One branch references a new *in situ* technology, while the other branch references excavation and disposal. The total estimated cost for installation, start-up, operation, maintenance, monitoring, and decommissioning of the new *in situ* technology is estimated to be $500,000 (if the technology achieves its intended goal). This amount is shown on the right-hand side and beneath the branch for new *in situ* technology. The estimated cost to excavate and dispose of the soil is $1,000,000 (see bottom right-hand side of branch). After meeting with the new technology vendor and a number of industry experts,

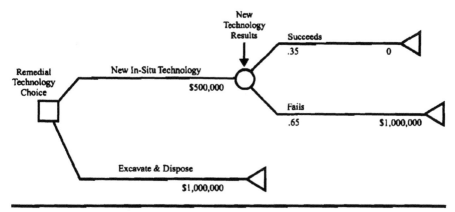

FIGURE 10.11 A basic risky decision.

the environmental manager concludes that there is a 35% probability that the new *in situ* technology will be effective at the site.

The tree is analyzed as shown in Figure 10.12. This figure indicates that there are three possible paths through the tree, designated A, B, and C. If the new technology is chosen and it achieves the intended cleanup goals, the project would be completed at a cost of $500,000 (path A). However, if the technology fails to achieve the intended cleanup goals, excavation and disposal will still be required, resulting in a total project cost of $1,500,000.

The expected monetary value associated with choosing the new technology is $1,150,000. This value is determined by summing the results of multiplying the end-point cost for each branch associated with the chance node for new technology results by the particular branch probability:

$$
\begin{aligned}
0.35 \times \quad \$500{,}000 &= \quad \$175{,}000 \\
+\ 0.65 \times \$1{,}500{,}000 &= \quad \$975{,}000 \\
&= \$1{,}150{,}000
\end{aligned}
$$

The expected monetary value associated with choosing the new technology is $150,000 greater than proceeding directly to excavation and disposal. Therefore, in this case, the recommended decision (i.e., the optimum decision policy) would be to proceed directly to excavation and disposal.

10.9.1.2 Decision Tree Demonstration of the Value of Information

The previous example provides a course of action based on an understanding of the information available at the time, in particular the estimated cost of the new

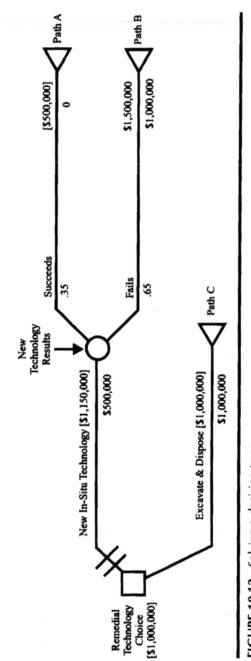

FIGURE 10.12 Solving a decision tree.

in situ technology and the likelihood that it will succeed. The optimum decision policy is a direct result of the assigned cost values and probabilities. Therefore, changes in any of these values will impact expected monetary values and, if significant enough, the optimum decision policy. Sensitivity analysis, as will be discussed later, can be utilized to understand the effect of probabilities and cost assignments on EMVs and the optimum decision policy. However, in this section, we will focus on a simple problem involving the value of information.

Suppose the new technology vendor indicates that, based on actual data, 85% of the sites that had positive pilot test results for the new technology achieved required cleanup standards. The vendor further indicates that, based on experience at other locations, there is a 50% probability that the pilot test would yield positive results at this site. The pilot test will cost $50,000. The environmental manager must now decide whether or not it makes sense to spend $50,000 on the pilot test. Figure 10.13 presents the analyzed decision tree structure under this new set of circumstances.

The analysis indicates that the optimum decision policy is to perform the pilot test and, if the test is successful, implement the new technology. In effect, the analysis indicates that investment in the pilot test will be money well spent because of its potential impact on our estimates of probability of success with the new technology.

The EMV associated with following the optimum decision policy is $875,000. Recall that this number is the probability weighted average. As an average, this value could never actually be realized. Another way of thinking about the EMV is that if this same project could be performed over and over again, sometimes ending in path A and other times in paths B and C, the average project cost would be $875,000. Because the EMV is an unbiased estimator, it is the best value to use when comparing alternative choices. The main alternative choice in this case is to proceed directly to excavation and disposal, which has an expected value outcome of $1,000,000 (which would have been the choice if we did not consider the pilot test). The costs that could be actualized if the optimum decision policy is followed are those associated with the end points of paths A ($550,000), B ($1,550,000), and D ($1,050,000), which have probabilities of 42.5, 7.5, and 50%, respectively (in accordance with Equation 10.4). Therefore, by investing in the pilot test, we are purchasing a 42.5% chance of completing the project for $550,000 and thereby saving $450,000, which if actualized represents a 5 to 1 savings.

It is possible to generate a cumulative probability versus cost curve for the optimum decision policy in Figure 10.13 (see Figure 10.14). This graph, also known as a risk profile, takes on a blocky appearance since there are only three possible project outcomes.

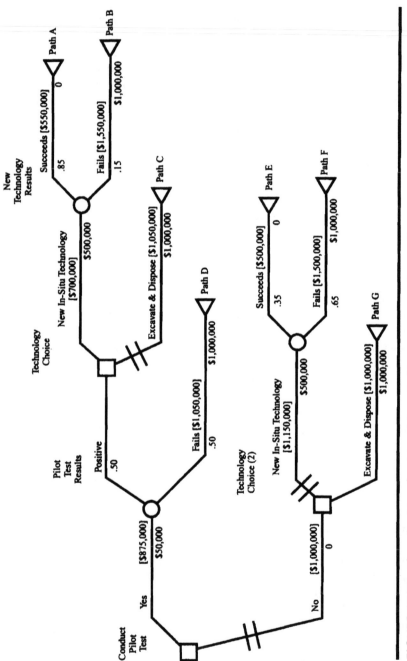

FIGURE 10.13 The solved tree for determining whether or not to conduct a pilot test.

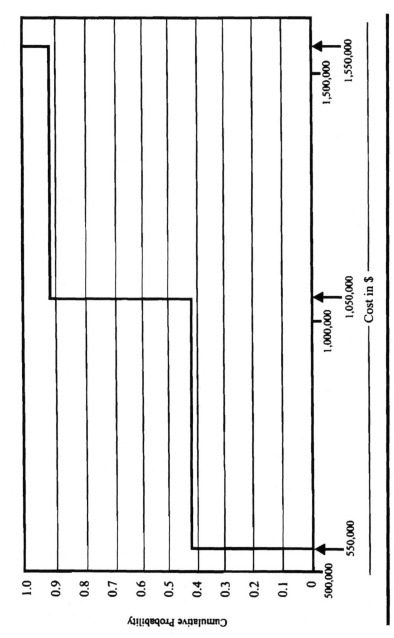

FIGURE 10.14 Cumulative probability distribution for the pilot test decision tree.

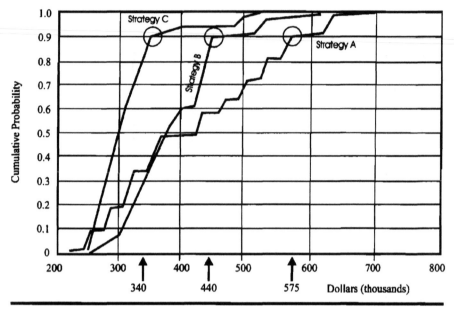

FIGURE 10.15 Comparing risk profiles.

If there are many chance events (as well as choices) after the original choice, the cumulative probability curve would take on a smoother, almost continuous looking profile. The shape of these profiles, in addition to expected values, can be utilized to compare alternative branches associated with the primary choice node. Figure 10.15 illustrates this point. The figure shows three different cumulative probability versus cost curves representing three different branches emanating from the initial choice node for an actual remediation project. Curve A represents the strategy that the site owner originally intended to follow prior to hiring an environmental consultant. Curve B represents the strategy recommended by the environmental consultant. Path C represents the curve associated with the optimum decision policy identified by decision tree analysis performed by the site owner and the consultant working together. The expected values associated with curves A, B, and C are $410,000, $390,000, and $316,000, respectively. Again, the expected values are probability weighted averages that can be utilized to compare and select decision alternatives. However, they do not tell the complete story. If we look at the shape of the three curves, we see that curve C is the steepest, which means that there is a lot less risk associated with this strategy than with the strategies associated with curves A and B. Curve C, which is steeper and to the left of curves B and C, is said to be stochastically superior.

Assume that the site owners have a policy whereby they want to be 90% certain that the budget they establish will not be exceeded. If they choose the strategy associated with curve C, they will have to set aside $340,000. On the other hand, if the strategies associated with curves A and B are followed, they will have to set aside a budget of $440,000 and $575,000, respectively.

10.9.1.3 Advantages and Disadvantages of Decision Trees

Now that we have constructed two basic decision trees, we can note some of the basic advantages and disadvantages of this technique. The first advantage is that the decision tree provides an extremely powerful and intuitive way to represent both the decision structure and the policy result.[18] A second advantage is that the decision tree makes the sequential timing (chronology) of the choices and chance events immediately obvious. The third advantage is that the tree provides clear pathways (outcome scenarios) that can easily be followed to any outcome cost.

The primary disadvantage of decision trees is that they can become very large very quickly, making them, from a practical standpoint, hard to print out and display.[19] Another disadvantage of decision trees is that they do not display probability and value dependencies well. For example, in Figure 10.13, it is not immediately clear that the probabilities associated with the new technology results are affected by the outcome of the pilot test. As we will soon see, this is immediately apparent using an influence diagram. In addition, referring again to Figure 10.13, there is nothing that indicates the component values which make up the cost values on the branches emanating from the choice and chance nodes.

10.9.2 Influence Diagrams

Influence diagram are similar to decision trees in that they provide a graphical representation of a decision problem. However, the representation provided an by influence diagram is usually more compact than that provided by a decision tree and better at communicating probability and value dependencies. Like decision trees, influence diagrams utilize squares to represent choice nodes and circles (more appropriately ovals) to represent chance nodes. The influence diagram also introduces a third node: the value node. Value nodes are represented by rectangles with rounded corners. These value nodes are equivalent to the values that are found on the right-hand side and beneath the branch lines emanating from the choice and chance nodes on the decision tree.

The various nodes on an influence diagram are connected by arrows, or arcs, which indicate relationships among the nodes. Similar to the construction of a network scheduling diagram, a node at the beginning of an arc is called a *pre-*

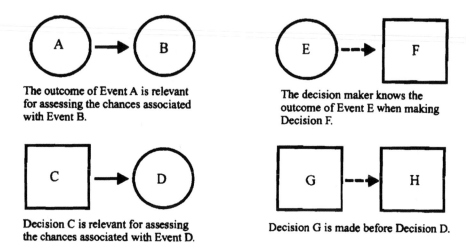

The outcome of Event A is relevant for assessing the chances associated with Event B.

The decision maker knows the outcome of Event E when making Decision F.

Decision C is relevant for assessing the chances associated with Event D.

Decision G is made before Decision D.

FIGURE 10.16 Representing influence with arrows. (From *Making Hard Decisions: An Introduction to Decision Analysis,* 2/e by R.T. Clemen. Copyright ©1996, 1991 Wadsworth Publishing Company. By permission of Brooks/Cole Publishing Company, Pacific Grove, CA, a division of International Thomson Publishing Inc.)

decessor and a node at the end of an arc is called a *successor.* The traditional rules for using arcs to represent relationships between chance and choice nodes are presented in Figure 10.16. The term traditional is used because new software programs permit additional, more flexible relationship arcs. However, for our purposes, there is no need to discuss these additional arcs.

With regard to Figure 10.16, it should be noted that no rules are provided for arcs leading into value nodes from either choice nodes or chance nodes. In general, arcs leading into value nodes from either choice nodes or chance nodes indicate that the value of the outcome depends on what happens (chance nodes) or what is decided (choice nodes) directly preceding the value node.[20] Arcs drawn from value nodes to other value nodes indicate that the predecessor value is a factor in calculating the value of the successor value. All arcs leading into or emanating from value nodes are traditionally represented as solid black lines.

10.9.2.1 Analyzing a Basic Risky Decision with an Influence Diagram

Figure 10.17 presents the influence diagram equivalent of the basic risky decision presented in decision tree format in Figure 10.11. Note that the choice and chance nodes do not have any branches emanating from them, as in the decision tree. This is because the information associated with these nodes is suppressed in order to better communicate the problem structure.

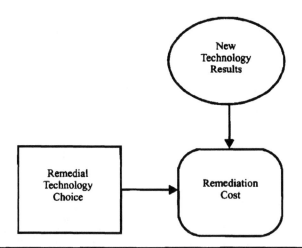

FIGURE 10.17   Influence diagram representation of a basic risky decision.

Figure 10.17 clearly indicates that the remediation cost depends on the remedial technology choice and the new technology results (if chosen). The fact that no arc proceeds from the remedial technology choice node to the new technology results chance node provides a clear indication that the technology choice has no bearing on the probability that the technology will result in success or failure.

In order to reveal more of the influence diagram's surface details, additional tables of information or (in some cases) decision trees are required.[21] Software programs designed for the purpose of creating and analyzing influence diagrams often have input windows which can be opened by the user to specify additional node information such as branch names, probabilities, and values. The suppressed data associated with Figure 10.17 are presented in Figure 10.18. Influence diagrams use the same general methods as decision trees. The solution to Figure 10.18 would be the same as the solution for Figure 10.11 (i.e., the optimum decision policy would be to proceed directly to excavation and disposal).

10.9.2.2   Influence Diagram Demonstration of the Value of Information

The influence diagram equivalent of the decision tree that was created to determined whether or not to perform the pilot test (Figure 10.12) is presented in Figure 10.19. Solving this diagram would lead to the same expected value, optimum decision policy, and risk profile as for Figure 10.12. The main difference is how the figure communicates the decision problem. This figure indicates that the decision to conduct the pilot test will be made before either of the

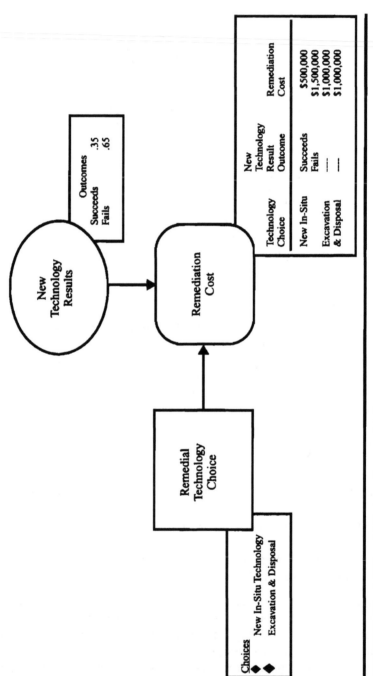

FIGURE 10.18 Additional influence diagram detail.

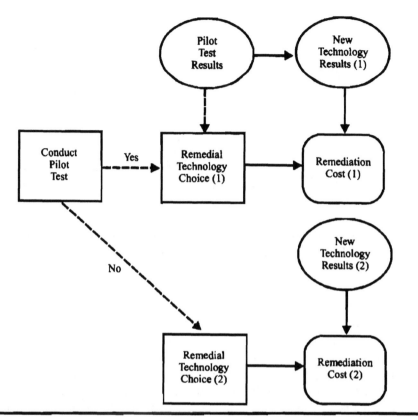

FIGURE 10.19 Influence diagram for determining whether or not to conduct a pilot test.

decisions designated as remedial technology (1) and remedial technology (2). These two nodes involve the same decision; however, as indicated by the influence arc leading from the chance node for pilot test results into the choice node for remedial technology (1), the results of the pilot test will be known before having to make the remedial technology choice. On the other hand, the diagram indicates that if the environmental manager does not conduct the pilot test, the remedial technology choice will be made without the benefit of such a test.

A more important aspect of Figure 10.17 is the communication of probability dependencies. Note the solid line emanating from the chance node for pilot test results to the chance node for new technology results. This line indicates that the outcome of the pilot test is relevant in determining the probability that the new technology will be successful at the site. The fact that the probabilities associated with one chance node influence the probabilities associated with another

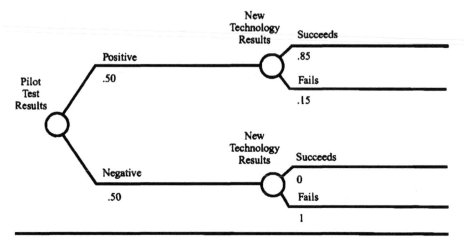

FIGURE 10.20 Influence of pilot test results on the probabilities for the new technology results.

chance node is clearly evident when using an influence diagram to model a decision problem.

Figure 10.19 was created by DPL™, a decision analysis program developed by Applied Decision Analysis, Inc. Opening the input window for the new technology results (1) chance node yields the diagram shown in Figure 10.20. This figure indicates that if the pilot test is successful, there is an 85% probability that the new technology will be successful at the site. On the other hand, if the pilot test is not successful, there is a 100% probability that the technology will fail at the site.

10.9.2.3 Value Dependencies and Influence Diagrams

As previously stated, one of the advantages of an influence diagram over a decision tree is the method by which it can be utilized to demonstrate value dependencies. When a decision tree is utilized, there is no indication of the components which make up the cost values provided on the tree. For example, in Figure 10.11, the decision tree for the original basic decision provides no information regarding the cost components of the new *in situ* technology. Such components would include the installation, start-up, monthly operations and maintenance (O&M), and decommissioning costs.

Influence diagrams provide the flexibility to indicate a variety of value components. These values can be in units other than dollars. To illustrate this concept, Figure 10.21 presents an expanded version of the influence diagram originally presented in Figure 10.19. This figure indicates that there are a number of

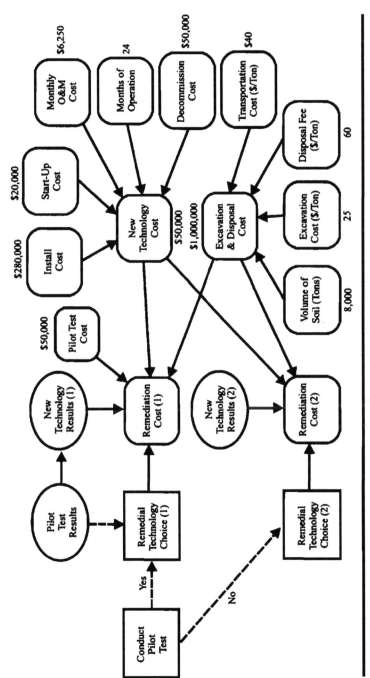

FIGURE 10.21 Expanded influence diagram showing additional value nodes.

component values that make up the two nodes for new technology cost and excavation and disposal cost.

The values nodes when included in the influence diagram in this fashion can be thought of a graphical representation of value cells that could be contained in a computer spreadsheet. Summary values such as the new technology cost node contain mathematical formulas (typical of spreadsheet formulas) which determine the value of a particular node based upon proper manipulation of input values. In the case of the new technology cost node, the input values include the installation, start-up, monthly O&M, and decommissioning costs, as well as the months of operation. Further investigation of Figure 10.21 indicates that remediation cost value nodes are influence by the new technology and the excavation and disposal value nodes. This is in addition to being influenced by the remedial technology choice nodes and the new technology results chance nodes. Finally, the value node for remediation cost (1) is also influenced by the pilot test cost. In solving the influence diagram, all possible output values for remediation costs 1 and 2 will be determined based on the input value nodes and all possible choice and chance states. Following the algorithm for influence diagrams and decision trees, expected monetary values for all chance events will be determined and higher EMV choices will be eliminated. The remaining choices leading to the overall lowest remediation cost will be included in the optimum decision policy. The solution in this case is the same as for the decision tree in Figure 10.11.

10.9.2.4 Value Sensitivity Analysis

Contained within the value node for new technology cost is a formula which sums up the installation, start-up, O&M (determined by multiplying the monthly cost by the total months of operation), and decommissioning costs. Sensitivity analysis is an effective technique for investigating the overall significance of these values on the optimum decision policy. For example, one of the more difficult values to estimate may be the number of months the *in situ* system will have to operate until the desired cleanup levels are achieved. The model currently uses a value of 24 months. Perhaps historical information, or simply expert judgment, indicates that the actual period of operation could be between 12 and 84 months. The longer the operation time, the higher the new technology cost and the greater the EMV associated with performing the pilot test. Figure 10.22 presents a sensitivity analysis of the operation time ranging from 12 months to 84 months. This is a graph of EMV cost (associated with the current decision policy) versus the months of system operation. Note that the EMV keeps increasing with increasing number of months until a value of 64 months is reached, at which point the curve breaks and flattens out. The break in the

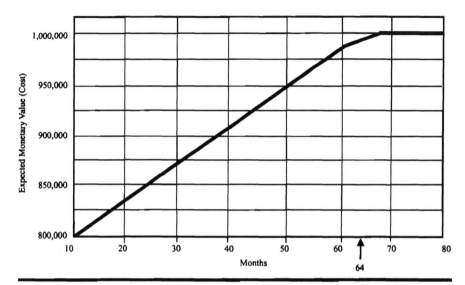

FIGURE 10.22   New technology operation time sensitivity analysis.

curve indicates a change in the optimum decision policy. The graph essentially indicates that if the environmental manager expects that the system would have to operate for more the 64 months (5.3 years) to achieve the intended cleanup goals, it would be better to proceed directly to excavation and disposal. In other words, the environmental manager could be off by nearly 40 months (original estimate 24 months) in estimating how long it will take to remediate the site and still be confident that the right decision was made in developing the strategy to perform the pilot test.

Another form of sensitivity analysis is to perform a sensitivity comparison of various input values. To perform such a comparison, a singular sensitivity analysis is performed on a number of input values, similar to the analysis of months of system operation. The results of the individual sensitivity analyses can be plotted on a single graph known as a tornado diagram, due to its characteristic appearance. Figure 10.23 is tornado diagram for a number of input values included in Figure 10.21.

The bar on this graph indicate the EMVs as a function of various parameter input values. The bars are ordered from top to bottom in order of decreasing impact on EMV (widest range to smallest range), which results in the characteristic tornado shape. The vertical line in the center of the graph represents the EMV ($875,000) for the optimum decision policy (perform pilot test). The bar at the top of the graph indicates that the volume of soil has been analyzed over a range of 5000 to 10,000 tons. This corresponds to a range in EMV of $625,000

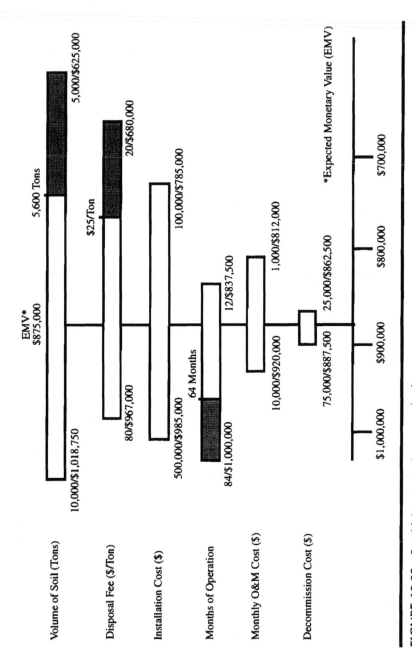

FIGURE 10.23 Sensitivity comparison tornado diagram.

to $1,018,750. On the top bar, there is a change in shading at a value of approximately 5600 tons. This shading indicates a change in decision policy. This indicates that if there is a high likelihood that the total volume of soil is below 5600 tons, it would be best to avoid the pilot test and proceed directly to excavation and disposal. The second bar indicates the same change in decision policy if it is believed that the soil could be disposed of for less than $25 per ton (assuming a volume of 8000 tons). The third bar indicates that the installation cost could go as high as $500,000 without affecting the optimum decision policy. The bottom two bars indicate that the monthly O&M and the decommissioning costs have little impact on the EMV and the optimum decision policy throughout their range of investigation.

The tornado diagram is extremely helpful in identifying significant parameters for the decision analysis. For those values that are most significant, the management team may decide that more information should be collected to eliminate some of the uncertainty associated with the base case value. The tornado diagram also indicates how robust the optimum decision policy is by helping the planning team determine how far off they can be in their estimates and still know they are making the right decisions.

10.9.2.5 Summary of the Advantages and Disadvantages of Influence Diagrams

The primary advantage of influence diagrams is that they provide an unambiguous representation of probabilistic and value dependencies.[22] Because of this, they are useful for creating easily understood overviews of decision problems. The directional arcs between uncertain variables clearly indicate dependence; the absence of arcs indicates independence.

A second advantage is that in decision problems where probabilistic dependence is an important feature, influence diagrams can graphically represent much larger models than decision trees, since each additional variable requires only a node and appropriate arcs, whereas trees cannot duplicate the same level of information without much more detail (i.e., many more branches).

The third important advantage of influence diagrams is that component values are easily indicated, which permits communication of the level of detail that went into determining higher level values. In addition, adding component values permits sensitivity analysis at a more detailed level.

10.9.3 Synthesis of Decision Trees and Influence Diagrams

The computer program DPL™, developed by Applied Decision Analysis, Inc., provides the ability to develop decision models using a synthesis of influence diagrams, decisions trees, and computer spreadsheets. These synthesized models overcome the disadvantages of traditional decision trees and influence diagrams by playing on the strengths of both. In the synthesis model, influence diagrams can be utilized to provide a clear indication of value and probabilistic dependencies. Component values can be exported to spreadsheet cells, where they are then functionally transformed by mathematical formulas contained in other spreadsheet cells and then exported to the influence diagram as higher level values. Finally, the sequence of decisions and choice and chance events is indicated by way of a decision tree. Programs such as this are extremely powerful, flexible, and capable of modeling the most complex of real-world problems. For the most part, they are limited only by the skill of the modeler.

10.10 MONTE CARLO SIMULATION

The previous sections dealt with the structuring of a decision problem using decision trees and influence diagrams to determine the optimum decision policy. In addition to providing the optimum decision policy, we saw that these techniques provide EMVs and cumulative probability distributions that can be utilized to better understand the likely project outcomes. In many decision analysis problems, the ultimate goal is the calculation of expected value.[23] However, in real-world problems, the number of uncertainty parameters can be so large that the decision tree will turn into a bushy mess or the influence diagram will become a maze of ovals and arcs. Although the synthesis approach discussed in the previous section can significantly help with this problem, in a variety of cases the Monte Carlo simulation may be more appropriate and easier to program.

The Monte Carlo simulation approach deals with complicated uncertainty models by considering the entire decision situation as an uncertain event. Legendary mathematician John Von Neuman has been credited with naming and popularizing the Monte Carlo technique while participating in the design of the atomic bomb.[24] Neuman recognized that a simple sampling technique can solve certain mathematical problems that are otherwise impossible.[24] Among the applications is solving for expected value, mean, or probability weighted average of a probability distribution. The term Monte Carlo simulation is often used to refer to this process because, as in gambling, the eventual results depend on random selection of the values of uncertain events.[25] The idea is straightforward: if the computer plays the game long enough, we will have a very good idea of the distribution of the possible results.[25] We can then analyze descriptive statistics dealing with both central location and dispersion of the data.

The basic method of Monte Carlo simulation is as follows:

1. Identify the structural relationship between uncertain variables. This can be a relatively simple mathematical function which relates the input values in a manner that determines the output value of interest.
2. Establish a probability density function for each input variable.
3. Form the cumulative distribution function for each variable by integrating the probability density function for each input variable.
4. Generate random numbers between 0 and 1 (the *y* axis of the cumulative distribution function) using a computerized random number generator.
5. For each random number generated, locate the corresponding variable value for each input parameter.
6. Calculate the output value of interest based on input values from the previous step and store the result.
7. Repeat steps 4 through 6 many times (usually more than 500 times).
8. Form the probability density function of the simulation results and analyze using descriptive statistics.

In theory, the Monte Carlo process can be utilized to model any decision problem that can be addressed by a decision tree or influence diagram. However, for our purposes, it is particularly useful in calculating the expected value of project duration. Recall that in Chapter 7, it was used to indicate that the time to complete a given task or project in general involves a high degree of uncertainty. Figure 10.24 provides further insight into the process steps and illustrates this technique as applied to project schedules. This figure indicates three project tasks linked in a finish-to-start fashion.

The duration of each task can be represented as a probability density function. Beneath each of the probability density functions, we see a corresponding cumulative density function for each task duration. Since all tasks are linked in a finish-to-start fashion, the total project duration is calculated by adding the three task durations. A random number generator is then utilized to identify the duration for each task. The sum of the task durations for a single run is stored, and the process is repeated over and over, which ultimately results in a probability density function for the entire project as well as descriptive statistics such as the mean project duration and the standard deviation.

10.11 PUTTING IT ALL TOGETHER: THE QDA/SPM PROCESS

Having studied the traditional project management planning techniques in Chapters 4 through 8 and the advanced techniques for evaluating uncertainty, we are

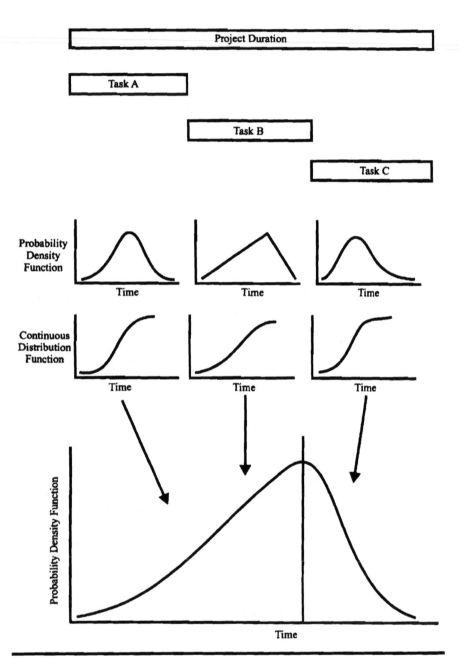

FIGURE 10.24 Monte Carlo simulation for determining expected value of project duration.

now in a position to put these techniques together, using all the techniques discussed in this book, in order to develop a comprehensive, multifaceted project plan in the face of uncertainty. The basic QDA/SPM process is as follows:

1. Perform data summary and stakeholder analysis (use the forms provided in Appendix B).
2. Identify project goals and objectives (with the aid of the summary/ stakeholder forms; also see Chapter 2).
3. Identify major decisions and chance events (use team planning process).
4. Construct skeleton decision tree/influence diagram (or synthesis).
5. Develop conceptual cost estimates and include in model (see Chapter 8 on cost estimating).
6. Solve for optimum decision policy; meet with planning team to discuss preliminary results.
7. Develop refined decision model.
8. Conduct sensitivity analysis; refine estimates of sensitive parameters.
9. Identify best, worst, and most likely paths through model.
10. Set up tasks for best, worst, and most likely project scenarios in a project management software package with simulation features (utilize the work breakdown structure templates in Appendix E and the guidelines for network scheduling provided in Chapter 7).
11. Perform Monte Carlo simulation of project time and construct cost over time curves for different scenarios.
12. Develop a report of advanced planning results.

The remainder of this chapter deals with the application of the above process to an actual project.

10.12   CASE STUDY EXAMPLE UTILIZING ADVANCED PLANNING TECHNIQUES*

This case study provides an example of using the techniques discussed in this book in support of a commercial property title transfer. Prior to offering the current tenant the opportunity to purchase the property, the owner decided to perform a modified Phase II and Phase III environmental assessment (as defined by the Pennsylvania Department of Environmental Protection [DEP]) and ana-

---

* This case study is drawn from an article entitled "Project Planning Use of Monte Carlo Simulations," by T.J. Havranek, published in *Remediation Management* (third quarter 1997, p. 1). It is reprinted here, with modification, by permission of *Remediation Management*.

lyze the data using the previously described advance planning techniques. The results of the quantitative decision analysis/project modeling allowed the owner to make sound business decisions with regard to the remedial action plan and to estimate the final cost/duration of remediation. This information led to significant cost savings and facilitated purchase negotiations.

Project Background Information

The case study involves a 2.4-acre property and an associated manufacturing/office building. At project initiation, the property and building were being leased to a cosmetics manufacturing company. The manufacturing company had expressed an interest in purchasing the property but was concerned about environmental impacts to the soil and groundwater which may have resulted from historical industrial operations. Therefore, the owner of the property was interested in determining the nature and extent of contamination (if any) and estimating the cost and duration associated with environmental remediation (if necessary).

Regulatory Climate

Shortly after project initiation (within two months), the state of Pennsylvania enacted new environmental legislation known as the Land Recycling and Environmental Remediation Standards Act (also referred to as Act 2 of 1995). Act 2 promotes the reuse of former industrial sites and safe, reasonable cleanups for existing business sites. The intent of this act is to resolve the technical and liability issues associated with former industrial property. These issues were widely believed to be negatively impacting the state's economic recovery. Of special interest, Act 2 provides a release from environmental liability to the owner, developer, purchaser, lending institution, and any other party participating in the cleanup of a site in accordance with the standards and procedures in the act.

Act 2 addresses past policies that set unnecessary, unreachable cleanup standards by clearly defining three options in terms of remediation standards for site cleanups. The three remediation standards are referred to as statewide health-based, background, and site-specific standards. The statewide health-based standards are human health-based standards for regulated substances in each environmental medium (soil and groundwater). Background standards are defined as concentrations of regulated substances present at a site but not related to a release at the site. Site-specific standards are established by way of a toxicological risk assessment which must be approved by the Pennsylvania DEP. Owners attaining either the statewide health-based standards or background standards are not subject to deed notification requirements. However, owners meeting site-specific standards are subject to deed notification requirements.

One of the decisions the site owners would face is which remediation standard to pursue. In order to make this decision, one would want to know which standard has the best chance of leading to the overall lowest project completion cost. Insight into this question can be provided by quantitative decision analysis.

Investigation Techniques and Results

A certain amount if information is necessary to develop a decision model that will closely reflect the problem at hand and not be overly complicated. For an environmental remediation project, the minimum amount of information is that which results from a typical Phase I investigation. With the Phase I data on hand, one can use QDA to identify the most cost-effective Phase II investigative strategy. Upon completion of the Phase II activities, QDA can be utilized to evaluate the most cost-effective remedial action strategy. Due to customer interests, the decision analysis was utilized for evaluation of the remedial action strategy on the case study project. Therefore, the investigation techniques and results obtained will be summarized briefly.

The investigative techniques involved:

- Site history data gathering
- Geoprobe shallow soil (<15 feet) gridded sampling
- Monitoring well installation and deep soil sampling (15 to 23 feet)
- Groundwater gauging, sampling
- Aquifer testing

The site history data gathering included client interviews, an environmental records search, and the collection of Sanborn maps and aerial photographs. The results obtained indicated that a petroleum refining company (gasoline and fuel oils) had been located directly north and east of the property during the early 1900s. In the 1930s, this company expanded operations onto the site in question by placing an office building and a small number of aboveground petroleum storage tanks (believed to have contained gasoline and fuel oil) on the property. In the late 1950s, the site was purchased by a bus transportation company, which in turn erected a bus garage (the current manufacturing building). In the late 1970s, the property was purchased by the current site owner, who in turn leased it to the cosmetics manufacturing company. The primary conclusion reached by the historical site characterization was that any environmental impacts found at the site would most likely be petroleum hydrocarbon related, primarily gasoline and diesel range organics. Therefore the list of analytical parameters included gasoline range organics (GROs), diesel range organics (DROs), and benzene, toluene, ethyl benzene, and xylene (BTEX).

The geoprobe soil sampling indicated GRO/DRO concentrations in excess of a 500-ppm cleanup standard that was in place prior to finalization of Act 2. The highest concentrations of these compounds were found in the south central portion of the site, extending to a depth of 14 feet (primarily DROs). With the aid of an iso-concentration map, it was estimated that there existed approximately 4000 cubic yards of soil with GRO/ DRO concentrations in excess of 500 ppm. Fortunately, none of the BTEX results in the soil were above the statewide health-based standards. In addition, the groundwater indicated nondetectable concentrations of all analytical parameters. Therefore, the remedial action plan would focus on the 4000 cubic yards of soil with GRO/DRO concentrations in excess of 500 ppm.

The finalization of Act 2 established statewide health-based standards for likely components of DROs, such as polynuclear aromatic hydrocarbons (PAHs). Additional sampling for PAHs would be required in order to evaluate where the site stood with regard to the statewide health-based standards. The locations with the highest concentrations of GROs/DROs could now be used to select the sampling locations for PAH analysis.

Technology Screening

As discussed in Chapter 2, over the past 10 to 15 years a number of technologies have been developed for the treatment of contaminated soil. In addition, depending on site conditions and soil properties, a wide variety of alternatives are available for how these technologies can be applied. In some cases, depending on the location, concentrations, and risks associated with the various chemicals, it may be acceptable and appropriate to perform little or no remediation. Due to the wide range of alternatives, a method of screening out some of the alternatives — prior to performing a detailed analysis using decision techniques — becomes necessary.

A discussion of the screening methodology would be quite lengthy and is beyond the scope of this case study. However, the method utilized was similar to that required for an RCRA Corrective Measures Study. The remediation alternatives remaining after application of the screening process included:

- Leaving existing levels in place
- *In situ* bioremediation
- *Ex situ* bioremediation
- Excavation and off-site disposal

The final cost of remediation cannot be estimated by simply choosing one of the above options. Leaving existing levels in place will require additional sampling to demonstrate levels below the health-based standards or a DEP-approved risk assessment indicating that site-specific standards have been met. The sampling for PAHs was estimated to cost $6300, while the estimated cost of the risk assessment was $15,000. It is possible that additional sampling will indicate that the PAHs are above the statewide health-based standards and that some form of remediation will be required.

The risk assessment might indicate that there are risks posed by the site and remediation would be required. In addition, even if the risk assessment indicates that the risks posed by the site are acceptable, there is a chance that the DEP will disagree with the risk assessment results and require some form of remediation. The risk assessment may conclude that the heath risks associated with the soil are acceptable if an asphalt cap is installed (estimated cost $83,400); again, the DEP could agree or disagree with this conclusion.

Both bioremediation technologies require testing to determine whether the site conditions will support these technologies. Bioremediation requires the presence of hydrocarbon-utilizing bacteria and no compounds in the soil which are toxic to such bacteria. The laboratory bench test needed to make this determination costs on the order $6500. This test may indicate that neither form of bioremediation (*in situ* or *ex situ*) will work at the site. *In situ* remediation was estimated at $130,200, while *ex situ* remediation was estimated at $203,300.

The site has a silty clay that appears at approximately 5 feet and extends to approximately 20 feet. This silty clay is not ideal for *in situ* bioremediation since it may not permit oxygen (the primary nutrient of bioremediation) to be drawn through the soil to a sufficient degree. However, the presence of this silty clay does not necessarily limit the application of this technology. A field pilot test is the best method for determining applicability. The estimated cost of the field pilot test is $10,300.

From the above discussion, it can be seen that there are a number of alternatives for addressing impacts at the site. Each option involves a cost and, depending on the results from the additional sampling, risk assessment, and bioremediation tests, new decision may be required regarding future actions to address site impacts. In order to estimate the most likely cost for completing the site cleanup, a logical method for identifying the best overall project strategy based on probable outcome of chance events is required. Determining the probabilities to assign to each chance event is the most subjective part of decision analysis and perhaps the most difficult. The choice of probabilities requires experience and insight; in many cases, one must draw on the opinions of industry experts. However, this subjectivity and initial apparent difficulty should not deter one from using this technique. The purpose of decision analysis is to help a decision maker think systematically about a complex problem and thereby improve the quality of the resulting decision. Any time spent thinking seriously about the subjective probabilities is likely to help the decision maker reach this goal.

Decision Tree Analysis

A decision tree for this site is presented as Figure 25. In this figure, the various decisions are represented by boxes and the chance events by circles. Costs associated with each decision node and chance event are located beneath and on the right-hand side of the line describing the event. The probabilities associated with each chance event are located beneath and on the left-hand side of the line describing the event.

The tree indicates that there are a number of pathways (branches) representing all possible decisions and chance events associated with addressing the impacted soil. According to the tree, the first decision is whether to perform the additional sampling or to proceed directly to site remediation. If sampling is chosen and indicates that the concentrations are above health-based standards, the site managers must decide whether to pursue a site-specific cleanup standard or statewide health-based standard. The former will require performing a risk assessment, while the latter will require some form of remediation. The risk assessment has three possible outcomes: acceptable risk (leave existing concentration in place), acceptable risk if an asphalt cap is installed, and unacceptable risk (remediate site). The acceptable risk scenarios will require DEP approval (see Figure 10.26). If the DEP does not approve of the risk assessment results, the site managers are left with remediating the site. If it is determined that site remediation is definitely required, the site managers are left with the choice of bioremediation or excavation/disposal. Each of the bioremediation options will require additional tests. If negative, these tests not only add to the overall project cost but also leave the site managers with only one option: off-site disposal (see Figure 10.27). If this happens, the site managers could have saved time and money by proceeding directly to off-site disposal without performing additional sampling, risk assessment, or bioremediation tests.

As previously discussed in this chapter, the tree is analyzed by moving from right to left and calculating the event status for each chance event. This is done by multiplying the probability by the cost for the various outcomes of the chance events. These values are then summed and reported as the event status. The cost for each test prior to the chance event is then added to the event status. This becomes the expected value for the chance event. At each decision point, the expected values for each decision are compared and the lowest cost decision is accepted. The higher cost decision is eliminated from further consideration.

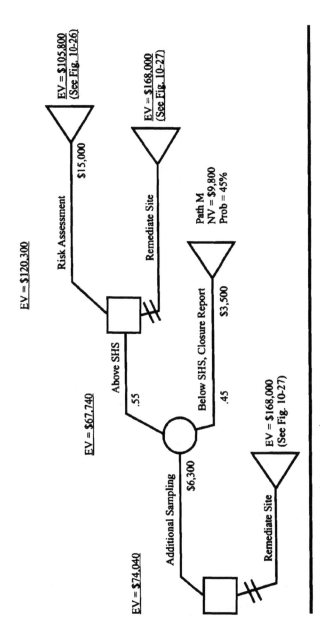

FIGURE 10.25 Remedial action plan.

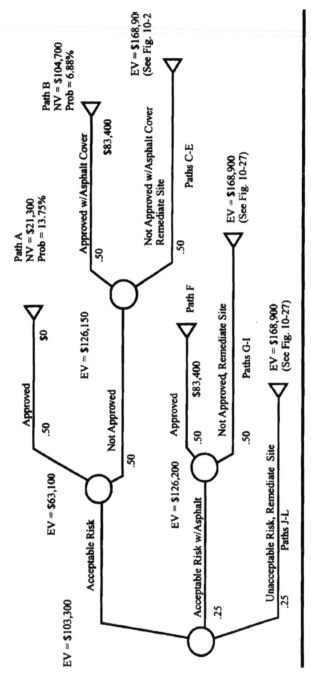

FIGURE 10.26 Risk assessment outcomes.

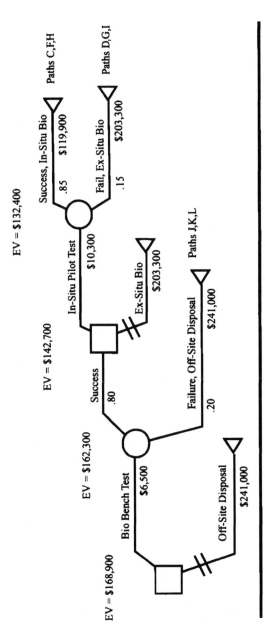

FIGURE 10.27 Remediation outcomes.

Proceeding in this fashion indicates that the best course of action would be to perform additional sampling at the site. If the results are above health-based standards, the next course of action is to perform a risk assessment. If the risk assessment indicates unacceptable concentrations or is not approved by the DEP, the next course of action is to propose an asphalt cap. If this does not achieve regulatory approval, the next best course of action is to perform a bench-scale bioremediation test. If the bench-scale bioremediation test is successful, the next best course of action is to pursue an *in situ* pilot test for bioremediation. Gaining this type of insight into a problem is often considered by many decision makers to be more important than the actual numerical results provided by the decision tree.

There are a total of 27 possible paths through the decision tree. Fourteen of the paths were eliminated because they were associated with decisions that would lead to excessive costs. Each of the 13 remaining paths has a probability of occurring. Given the number of possible outcomes, one can see why it is difficult without the aid of this technique for project managers to predict the final cost of an environmental project.

The 13 remaining paths are utilized to calculate the expected value for the entire project given the assigned costs and chance scenarios. The calculated expected value is $74,000. The expected value is the probability weighted average of all possible outcomes of the project. One way of viewing this number is that if we could execute this same project many times, the average cost per project would be $74,000. Of course, we will only execute the project once. The cost of the project when we are finished will be one of the net values calculated for the chain of decisions and chance events which lead to that particular outcome.

The calculated net values for our project range from a low of $9800 (path M) to a high of $268,000 (paths J, K, and L). Each of these paths has a calculated probability of occurring which is determined by multiplying all probabilities along the path moving from right to left. The least cost path (path M) has a probability of 45%, while the highest cost paths have a combined probability of 5.5%. Proceeding in this fashion, a schedule of cost versus probability can be developed for this project (provided below) and can be graphed as in Figure 10.28.

- Less than or equal to $9,800   45% probability
- Less than or equal to $21,000   59% probability
- Less than or equal to $104,700   73% probability
- Less than or equal to $158,000   92% probability
- Less than or equal to $241,400   95% probability
- Less than or equal to $268,000   99% probability

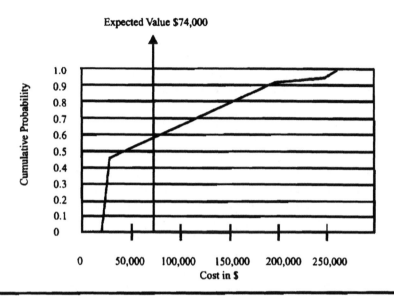

FIGURE 10.28   Cumulative probability versus cost curve for the case study project.

Up to now, our decision policy has been based on minimizing cost, and we have not considered the effective of our decisions on the duration of the project. If time were our primary concern, our decisions might have been significantly different. For example, the most costly remediation technique, excavation and disposal, has a relatively short treatment time.

It is possible to analyze the tree for project duration based on our current decision policy. Instead of cost, we must include an estimate of the amount of time needed to perform each task. In addition, we must remember to include the time necessary to complete tasks performed by outside parties, such as the amount of time for regulatory review and approval. Analysis of the tree in this fashion yielded an expected value of 200 working days (approximately 10 months). The analysis also indicated that there was a 62% chance of finishing the project in 95 days and a 95% chance of finishing the project in 505 working days (approximately two years).

Monte Carlo Simulation for Project Duration

The current owner of the site would be responsible for performing any remediation necessary until release from environmental liability could be obtained. Since it was believed that there was more uncertainty associated with the duration estimates than the cost estimates, a more in-depth analysis of project duration was performed. Stochastic project modeling (Monte Carlo simulation) was called on to assist in this task.

We begin by identifying the path with the longest expected duration by inspection. This would be the path where the customer made all the best decisions and had moderate to poor luck, eventually leading to implementation of *in situ* bioremediation (path C).

To perform this analysis, a work breakdown structure was developed for path C, including all possible tasks, even those performed by the regulatory agency. Instead of a single duration assigned to each task, a probability distribution representing the time to complete each task was input into the project model. Each task was then linked using network diagramming techniques. A Monte Carlo simulation was performed whereby trial values were sampled from each task distribution using a random number generator and then the overall project duration was calculated in light of the network diagram. This process was repeated 500 times, with the output from each iteration utilized to generate a probability distribution for overall project duration.

This simulation indicated that the expected time frame to complete the project is 555 days with a standard deviation of 36 days. The model also indicated that there is only a 15% chance of finishing path C in less than 505 days (the 95th percentile value from the decision tree analysis) and a 95% chance of finishing the project in less than 609 days (approximately 2.5 years).

Finally, a similar analysis was run for the "best-case" path (path M). The analysis indicated an expected value of 90 working days with a standard deviation of 8 days. The 10th percentile value was reported at 79 days and the 95th percentile value was reported at 103 days. A time-scaled network diagram for this path indicated that if the project began on November 13, 1996, it would be completed by March 12, 1996.

Project Outcome

The optimal decision policy was incorporated into a strategic plan that was submitted to the Pennsylvania DEP. The site owner decided to hold off on sale negotiations until the results from the additional sampling were obtained and submitted to the DEP. The PAH sampling results were below the statewide health-based standards. A final report was submitted to the regulatory agency along with a request for release from environmental liability. The costs for the additional sampling were within the expected budget of $6300 and $3500, respectively ($9800 total). A letter indicating release from environmental liability was received from the DEP on Mach 8, 1996. The sale was negotiated shortly after receipt of this letter.

These techniques allowed the site owner to make a sound business decision, hopefully leading to best-case scenarios, while at the same time being prepared for worst-case scenarios. The owner thereby replaced reactive remediation project planning with proactive, sound business management, ensuring both regulatory compliance and protection of the environment.

REFERENCES

1. Lindeburg, M.R., *Engineer-in-Training Reference Manual*, Professional Publications, Belmont, CA, 1992, 56-1.
2. Wideman, R.M., *Project and Program Risk Management: A Guide to Managing Project Risks and Opportunities*, Project Management Institute, Drexel Hill, PA, 1992, 1-3.
3. ADA Decision Systems, *DPL Standard Version User Guide*, Duxbury Press, Belmont, CA, 1995, 2.
4. Clemen, R.T., *Making Hard Decisions: An Introduction to Decision Analysis*, Duxbury Press, Belmont, CA, 1990, 4.
5. Clemen, R.T., *Making Hard Decisions: An Introduction to Decision Analysis*, Duxbury Press, Belmont, CA. 1990, 2.
6. Lindeburgh, M.R., *Engineer-in-Training Reference Manual*, Professional Publications, Belmont, CA, 1992, 22-1.
7. Fuller, R.B., *Synergetics, Explorations in the Geometry of Thinking*, MacMillian, New York, 1992, 95.
8. Edmondson, A.C., *A Fuller Explanation, The Synergetic Geometry of R. Buckminster Fuller*, Birkhauser Boston, Cambridge, MA, 1987, 31.
9. Edmondson, A.C., *A Fuller Explanation, The Synergetic Geometry of R. Buckminster Fuller*, Birkhauser Boston, Cambridge, MA, 1987, 32.
10. Clemen, R.T., *Making Hard Decisions: An Introduction to Decision Analysis*, Duxbury Press, Belmont, CA, 1990, 169.
11. Lindeburgh, M.R., *Engineer-in-Training Reference Manual*, Professional Publications, Belmont, CA, 1992, 22-1.
12. Schyyler, J.R., *Decision Analysis in Projects*, Project Management Institute, Upper Darby, PA, 1996, 5.
13. Lindeburgh, M.R., *Engineer-in-Training Reference Manual*, Professional Publications, Belmont, CA, 1992, 11-6.
14. Lindeburgh, M.R., *Engineer-in-Training Reference Manual*, Professional Publications, Belmont, 1992, CA, 11-9.
15. Clemen, R.T., *Making Hard Decisions: An Introduction to Decision Analysis*, Duxbury Press, Belmont, CA, 1990, 49.
16. ADA Decision Systems, *DPL Standard Version User Guide*, Duxbury Press, Belmont, CA, 1995, 224.
17. Clemen, R.T., *Making Hard Decisions: An Introduction to Decision Analysis*, Duxbury Press, Belmont, CA, 1990, 34.
18. ADA Decision Systems, *DPL Standard Version User Guide*, Duxbury Press, Belmont, CA, 1995, 225.

19. ADA Decision Systems, *DPL Standard Version User Guide,* Duxbury Press, Belmont, CA, 1995, 228.

20. Clemen, R.T., *Making Hard Decisions: An Introduction to Decision Analysis,* Duxbury Press, Belmont, CA, 1990, 35.

21. Clemen, R.T., *Making Hard Decisions: An Introduction to Decision Analysis,* Duxbury Press, Belmont, CA, 1990, 21.

22. ADA Decision Systems, *DPL Standard Version User Guide,* Duxbury Press, Belmont, CA, 1995, 226.

23. Clemen, R.T., *Making Hard Decisions: An Introduction to Decision Analysis,* Duxbury Press, Belmont, CA, 1990, 313.

24. Schyyler, J.R., *Decision Analysis in Projects,* Project Management Institute, Upper Darby, PA, 1996, 44.

25. Clemen, R.T., *Making Hard Decisions: An Introduction to Decision Analysis,* Duxbury Press, Belmont, CA, 1990, 316.

26. Havranek, T.J., Project management use of Monte Carlo simulations, *Remediation Management,* Third Quarter, 13, 1997.

Implementing Modern Project Management

11

It is evident from the preceding chapters that modern project management techniques can significantly improve environmental remediation project performance, in terms of lower cost and shorter duration, while at the same time protecting human health and the environment. These improvements go a long way toward serving the interests of site owners and the public in general. In addition, the improved performance gained by the implementation of modern project management processes also serves the interests of environmental service companies. The efficiencies achieved through the implementation of modern project management permit increased profit margins even though the total cost to the customer is less. This is in accord with an old business adage: *It's not how much you make, it's how much you keep.*

With the benefits of modern project management techniques understood, the question for any company wishing to improve project performance becomes: *How do we create the proper environment and support systems to take full advantage of these techniques?* It must be understood that without the proper support systems in place, it will be impossible for the project managers to make full use of modern project management techniques or for the organization at large to realize the benefits. This chapter discusses the support systems that must be in place to facilitate modern project management. It provides recommendations for implementing modern project management by way of a case study example of an actual implementation project. This case study also demonstrates some of the difficulties experienced by a company in moving toward full modern project management implementation. The experience is common among many companies that have recognized the need for modern project management but are struggling to find the most effective way to implement the changes.

Hopefully, readers can draw on these experiences and avoid some of the difficulties. Finally, it should be understood that this chapter is not written only for environmental service companies; it provides useful information for any company wishing to improve project management processes.

11.1 THE PROJECT PLANNING AND CONTROL SYSTEM

To maintain a competitive edge as well as a favorable bottom line, many organizations have found it necessary to develop a project planning and control system (PP&CS) which fully supports their style of managing projects. Simply stated, a PP&CS includes the process, information system, and trained personnel necessary for effective project planning and control.[1]

The components of the PP&CS are illustrated in Figure 11.1. The *process* represents the overall methods and procedures by which projects are planned and controlled. The previous chapters provided a standardized project planning process and tools specific to the environmental remediation industry. The process and tools provided can be used to lay the foundation for the PP&CS within any environmental service company (or any company for that matter). Note that the advanced project planning techniques presented in Chapter 10 fit into the overall planning process at the stage of determining project goals and objectives.

Another way of viewing the process is that it represents the way an organization "does business" and therefore is the foundation of the PP&CS. The other components (information system and trained personnel) facilitate implementation of this process by providing the framework of the PP&CS. Figure 11.2 illustrates the process that forms the foundation and the other components that form the framework of the PP&CS.

The *information system* encompasses all computerized tools and associated interfaces necessary to sustain the process. A proper information system is one that, at a minimum, links an appropriate project management software program

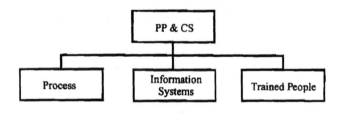

FIGURE 11.1 Components of the project planning and control system. (From Sam Provil, PMP—Project Controls, Inc. With permission.)

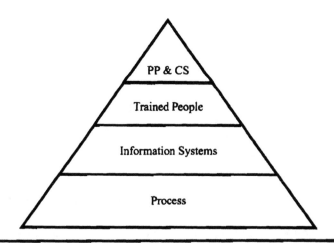

FIGURE 11.2 Process: the foundation of the project planning and control system framework. (From Sam Provil, PMP—Project Controls, Inc. With permission.)

with a company's financial accounting system. The details of creating such a system are presented in the upcoming case study.

Effective use of a PP&CS requires *people* — who are trained and skilled in both the process of modern project management and use of the information system. It should be understood that it is not enough to train only the project managers. For modern project management to work, it must become part of a company's culture.

The remainder of this chapter presents a case study implementation of modern project management in a major environmental service company. The case study originally appeared in a paper entitled "Implementing Modern Project Management in an Environmental Service Company," by Havranek et. al., published in the *Proceedings of the 27th Annual Seminar/Symposium of the Project Management Institute.*[3] It is reprinted, with modification, by permission of the Project Management Institute.

11.2 SYNERGETIC APPROACH TO MODERN PROJECT MANAGEMENT IMPLEMENTATION

Perhaps one of the first things to bear in mind when attempting to implement modern project management in any company is that it is best done by a team of individuals representative of the company's operational groups. The implementation itself must be viewed as a project. Therefore it will require all the elements of a good project plan, such as identified goals and objectives, project team, scope, schedule, and budget. Reengineering a company to make full use of modern project management processes can require anywhere from six months to two years depending on the size of the organization and current level of project management sophistication.

The case study project team was able to draw on historical attempts by the organization to improve project management. Drawing on these historical attempts and researching established industry practice, the modern project management implementation team settled on a comprehensive, synergistic approach which included the following elements:

- The development of an industry/company-specific training module based on the Project Management Institute's *Project Management Body of Knowledge* and well-established authors in the field of project management
- The development of a project management information system, accessible over a wide area network, that links the company's financial accounting software and its selected project management program
- Localized in-office training in modern project management theory and the use of the project management information system

For the above elements to be effective, they must be carried out in consideration of the culture of a company and the industry in which it operates. Therefore, before presenting the techniques employed in the case study, the following section discusses the history of the company and its industry and relates them to the concept of the industry growth cycle.

11.3   THE INDUSTRY GROWTH CYCLE

According to Singh and Linck, an industry begins due to some form of innovation or need.[2] Once the need is recognized, there is a period of growth/boom in the industry. Next, this rapid growth creates problems, such as finding skilled personnel and managing a larger organization.[2] As the industry matures, competition arises and profits are reduced. Finally, the growing industry requires improved performance.[2] It is at this point that successful companies in many industries have achieved desired performance through the implementation of modern project management.

The experience of the case study company validates the industry growth cycle. When the company was founded in the late 1970s, innovation was required to address new environmental regulations. The company responded through the development of a number of remedial technology innovations.

As a result of these developments, the company gained early industry recognition. During the mid-1980s, the case study company experienced rapid growth, as did other companies that continued to push the envelope by using the latest advances in environmental remediation.

During the late 1980s, growth problems arose in managing the expanding organization. In less than one decade, the company had grown from a small start-up to more than 2000 people, with offices located throughout the United States.

Competition increased during the early 1990s as more firms entered the market and customers became more sophisticated. Efforts to meet customer demands and maintain a competitive edge cut into profit margins. Having reached this need for improved performance, the company began to look for a method for improving project management.

11.4   HISTORICAL ATTEMPTS AT MODERN PROJECT MANAGEMENT IMPLEMENTATION

Stage 1: Outside consultant, project management fundamentals.
Management's first attempt at improved project performance was through training. In 1991, an outside consultant was hired to put on a series of week-long training sessions on project management fundamentals. Although many employees commented that they found the sessions informative, only a handful were actually utilizing these techniques in their daily work. Many employees cited difficulties in moving from theory to practical application, while others cited a lack of equipment and software support systems. Others indicated that their customers were content with the current level of project performance and reluctant to pay for additional project management controls. Ultimately, the training had little impact on project performance.

Stage 2: In-house training, pragmatic project management. Drawing on the lessons of the previous program, the company decided to develop an in-house project management training program. Senior employees who were successful in managing some of the company's larger projects were assigned to this task. The overall theme of this training was the implementation of pragmatic management techniques. The development team began by defining a series of project manager responsibilities. A case study project was employed to demonstrate the project manager's responsibilities. Fundamental concepts such as work breakdown structure and network scheduling were avoided. This program was implemented in 1992 and, unfortunately, it too had little impact on project performance. In addition, many of the individuals who received the training were overwhelmed by the host of new responsibilities without any support systems to carry them out.

Stage 3: Project management procedures manual. With the limited results from the two previous programs, it was concluded that performance might be improved by mandating a series of standardized project management procedures. In others words, attempts were made to define a standardized project management process. A procedures manual was distributed throughout the company. Audits were planned to ensure that these procedures were being followed. The audits were implemented, but only to a limited degree due to time and billability pressures. As a result, these procedures were not adopted on a large scale.

11.5 DEVELOPMENT OF THE MODERN PROJECT MANAGEMENT IMPLEMENTATION PROJECT

The historical attempts at modern project management implementation indicated that success could not be obtained by "band-aid" measures. The design solution would have to be comprehensive in scope, and synergetic in application, while drawing on the lessons learned from the past. With this in mind, the project team for the implementation project began to develop a series of goals and objectives which, if obtained, would be in support of the company's corporate mission.

11.5.1 Goals and Objectives

The overall objective of the project was the implementation of a project management process fully automated within a software environment and supported by trained personnel. The subobjectives of the implementation included:

- Enhancing the planning capabilities of the project managers
- Providing project managers with practical planning processes and tools specific to the environmental industry
- Increasing awareness and understanding of modern project management methods and tools at all levels throughout the company

11.5.2 Conceptual Approach

The following lessons learned were considered in the design of the conceptual approach for the modern project management implementation project:

- Training that is either overly theoretical or pragmatic will not achieve long-term results. Project managers have difficulty with practical application of the former and coping with the inflexibility of the latter.
- Training should be specific to the industry and the company in light of best available practices as promoted by recognized leaders in project management theory and organizations such as the Project Management Institute.
- Mandating a series of responsibilities and procedures without providing support systems to see that they are met overburdens project managers without achieving effective results. "Tools, Not Rules" would become the guiding theme.
- The training program and the support systems should be developed and implemented concurrently to be certain that there is agreement between what is being taught and what can be carried out in practice.
- Training should be provided to a wide spectrum of employees, including regional managers, district managers, task managers, and finance managers, in addition to project managers and project management assistants, to ensure the modern project management techniques become ingrained in the corporate culture.
- The training program should be developed by an outside consultant working closely with an internal team to address pragmatic concerns while incorporating the best of project management theory.

The primary tasks of implementing modern project management in the company were to develop an industry-relevant project management training module and a project management information system that was accessible over the company's wide area network. This would be followed by a test location rollout. The final task would be companywide rollout.

11.6 DEVELOPMENT OF THE TRAINING MODULE

Development of the training module was contracted to an outside service company. The company sought a training program that would incorporate the practical demands of the environmental remediation industry into the best of modern project management theory. A five-member internal team was assigned to comment on and review all aspects of the training module development.

The training module focused on project planning since all involved agreed that sound project management begins with quality project planning. The heart of the training module would be a standardized planning process that could be applied to all types of remediation projects. This process is similar to the one presented in Chapter 3 and supported by Chapters 4 through 9.

11.7 DEVELOPMENT OF THE PROJECT MANAGEMENT INFORMATION SYSTEM

The case study company investigated a number of project management software programs and benchmarked several products. The company's technical services department led the project management system development in a team effort with the company's MIS department. The technical services department profiled the customer base of various project management software vendors, looking for products that had excelled in companies comparable to the case study company. The case study company is a widely geographically distributed organization, with a large number of small offices and a large number of small to medium-sized projects. After reviewing the options, the technical services department selected Artemis Prestige™, a multiuser, multiproject management system optimized for the client/server environment.

It should be noted that from a project management perspective, many of the products investigated could have fulfilled the requirements, but the key differentiator for the Artemis product was the capability for implementation on a companywide scale, with all users accessing the same central database. The company did not want to face the prospect of running multiple databases and multiple extracts and updates from the accounting system, each creating the potential for process failure and lack of assurance of data integrity. Microsoft Access™ was used as a database query tool to leverage the defined tabled structure of Artemis Prestige™ to create custom reports.

11.7.1 Oracle™ Database

The case study company stores project information in an Oracle™ relational database management system. Important evaluation criteria that drove the selection of Oracle™ included the product's reputation, the size of the installed base, the depth of the product support staff, and the level of announced and planned upgrades for the product. The intention was that as the data warehouse grows, the standard data replication/distribution functionality contained within Oracle™ will enable the dissemination and synchronization of project (relational) data over multiple servers in several major offices. The multidimensional (spatial) data capabilities contained within the most recent release of Oracle™ enable the integration and analysis of both project data and site data. This function would normally require the manual integration of project-based information with site data stored in a separate Geographical Information System (GIS) tool. Over time, it is expected that this will allow the company to seamlessly integrate both relational and spatial data (i.e., the comparison of contaminant levels at a specific site to the technologies as well as the costs generated by a particular remedial strategy) into a consolidated decision support environment.

The company collects cost information from a mainframe legacy accounting system in a data warehouse environment which contains over 100 Oracle tables/views and uses Microsoft Access™ to integrate these data with the project setup and scheduling information entered through Artemis Prestige™ (see Figure 11.3). Thus, the company is able to develop proposals with the benefit of historical actuals — down to a project/activity/resource level. Microsoft Access™ is also used in concert with Artemis Prestige™ to provide a wide range of function-specific modules that support the different phases of the overall project life cycle

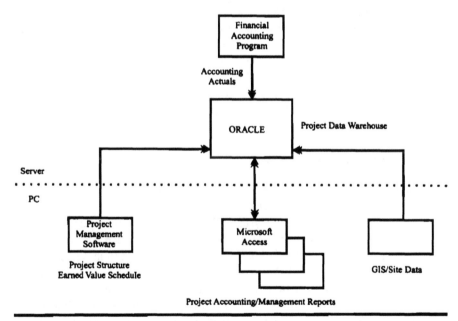

FIGURE 11.3   Program schematic. (From Havranek, T.J., Lavin, W.M., and Sullivan, M.H., Implementing modern project management in an environmental service company, in *Proceedings of the 27th Annual Seminar/Symposium,* Project Management Institute, Drexel Hill, PA, 1992, 248. Reprinted with permission of the Project Management Institute, 130 South State Road, Upper Darby, PA 19082, a worldwide organization of advancing the state-of-the-art in project management. Phone: (610) 734-3330, fax: (610) 734-3266.)

(i.e., as an estimating tool, marrying the cost and project information to enable the development of accurate forecasts).

11.7.2 User-Friendly Data Management and Report Writer

Microsoft Access™ provided a powerful tool to support the Rapid Application Development (RAD) utilized by the company. The use of various Template Wizards available within Microsoft Access™ allowed for the quick prototyping and evaluation of alternative end-user solutions.

Microsoft Access™, through its Open Database Connectivity (ODBC), allows users to keep customer- and/or site-specific information in local databases and marry that information with project data in the data warehouse. Access reports were developed to combine the scheduling and basic project information set up in Prestige™ with the actual project charges that were downloaded from the accounting system each week. Several internal and customer-focused monitoring and control reports were developed, in particular to enable earned value analysis. Other modern project management modules that have been developed include work order management, resource management, and rate table maintenance, as well as multiproject (program) management.

Finally, the company also used Microsoft Access™ to create all of its user interfaces and to generate most of its project-level reports. The development of custom interface screens familiar to the users enabled training to progress rapidly.

11.8 ON-SITE TRAINING

The modern project management implementation project for our case study company employed concurrent training in project planning and the use of the project management information system. This technique of providing training in the planning concepts and the use of support tools was considered key to modern project management implementation. It allowed the trainees to make immediate use of their newly developed skills by applying them with the custom-designed project management information system. Registered participants were provided with a case study project summary in advance of the workshop (similar to that provided in Appendix A). This maximized the time spent in group break-out sessions. During the break-out sessions, the attendees were given the opportunity to enter relevant project data into the project management information system.

The general format of the training workshop consisted of a series of one-hour lectures followed by case study exercises designed to permit practical application of the tools and techniques discussed in each lecture. The workshop was offered over a three-day period. Two and one-half days was required for lecture and class exercises. The remaining half-day was available for participants to work with the instructors on detailed systems training and to discuss how the project process will be carried out in the office.

11.8.1 Training Exit Surveys

Approximately 250 employees attended the training seminars, and the exit surveys for the training were overwhelmingly positive. The attendees gave high ratings to all aspects of the training and indicated that rolling out the system concurrent with the training was a highly effective approach.

11.8.2 Project Performance Changes

The primary project performance changes are in the areas of:

- Improved customer communication
- Improved project scope development, cost estimating, and scheduling
- Increased profitability and the ability to project profitability
- Ability to predict future resource needs

Customer communication has been enhanced by the ability to better track project performance by using earned value techniques. Earned value statistics can now be provided with other billing and invoice information. Many customers reported increased confidence in the company's ability to manage technology implementation.

Project planning continues to improve due to the historical database of projects that can be accessed online. Profitability has been enhanced through decreased write-offs of unaccounted for time and better ability to identify and control project changes on a timely basis. Online access to all project plans has allowed operations managers to predict future resource needs. A final benefit is that project managers now have profitability objectives for their projects before they start and can see how their well-managed projects integrate with company profit objectives, thereby increasing their morale and their better understanding of their roles within the organization.

REFERENCES

1. Provil S., Process: the foundation of project planning and control systems, *Project Management News*, Vol. 94, No. 1, 1, 1994.

2. Singh, B.A. and Linck, C.J., Project management advancement program: getting maximum returns for environmental dollars, in *Proceedings Project Management Institute's 1992 Annual Seminar/Symposium*, Project Management Institute, Drexel Hill, PA, 1992, 134.

3. Havranek, T.J., Lavin, W.M., and Sullivan, M.H., Implementing modern project management in an environmental service company, in *Project Management Institute 1996 Proceedings, Revolutions, Evolutions, Project Solutions, 27th Annual Seminar/Symposium*, Project Management Institute, Drexel Hill, PA, 1996, 248.

Conclusion

12

In the beginning chapters of this book, we came to understand that although the environmental remediation industry is rather young in terms of years, it has rapidly gone through the industrial growth cycle, whereby customer demands, the competitive market, and the interests of the general public have created the need for improved project performance. We also came to understand that as the industry matured, little changed to reduce the level of uncertainty and hence the risks associated with environmental remediation projects.

The techniques of modern project management have been proposed as a solution to improve project performance. We took an in-depth look at project management theory and identified a standardized project planning process. After applying the process, we came to recognize that uncertainty still remains in our project plans. In light of this recognition, advanced techniques for understanding and managing risks within the project plan were presented and demonstrated. These advanced techniques could have been presented near the beginning of the book as they assist in better defining project goals and objectives and evaluating alternative strategies and provide a steering mechanism for the overall project plan. However, had these techniques been presented early on, much confusion might have been created and their value underappreciated. To achieve the full benefit of the advanced project planning techniques, one must be sufficiently skilled in the other project management processes.

The number of pages that it took to present the various planning techniques, traditional and advanced, should not be viewed as an indicator of the amount of time or level of effort required to apply these techniques to actual projects. Once the process, tools, and techniques are understood, and the support systems for their use are in place, projects managers will find that multifaceted, comprehensive, high-quality project plans can be developed in much less time than with the methods they may currently be using.

Finally, it should be understood that even though the focus of this book has been on the environmental remediation industry, the techniques apply to all industries. Companies and individuals that become adept in their application will find that they can apply them in other industries in need of improved performance and reap the associated rewards.

Megacorp Site Summary

1 INTRODUCTION

This site summary provides information needed for exercises found at the end of Chapters 4 through 8. The information contained in this summary is based on an actual environmental remediation project. The circumstances of the site (e.g., customer name, extent of impact, location, etc.) have been altered, as have the circumstances of the service company attempting to respond to the project (e.g., name, resources available, nearest office location). This was done to protect the anonymity of all parties involved. In addition, some of the politics of the project, in terms of customer requirements, personalities, etc., have been have altered and exaggerated in order to make the exercises more interesting and challenging.

This site summary was developed to demonstrate project management principles and, therefore, is not intended as a lesson in site investigation techniques or technology selection. Technical information is provided only to the degree believed necessary to permit project planning.

The intention of this summary is to provide information in a manner similar to the way it is acquired during the concept phase of any environmental project. Therefore, the information may not be fully organized, structured, or complete. The organizing and structuring of the information will be the result of performing the exercises found at the end of the chapters. If it appears that information necessary to complete a given exercise is missing, it is acceptable to make assumptions. However, such assumptions should be documented for later project planning exercises.

2 THE CUSTOMER

Mega Manufacturing Corporation (Megacorp) is a Fortune 500 company with manufacturing plants located throughout the United States and overseas. Megacorp is attempting to proactively address environmental impacts at its various manufacturing plants. Its goal is to reduce, and make more accurate, the estimates of environmental liabilities that are supplied annually to its financial department.

The company has prioritized plants that are no longer in service and are planned to be sold. In addition, the company is focusing on plants that it no longer owns, with the exception of environmental liabilities retained as part of the sales agreement. As a result of its proactive policies, Megacorp is often ahead of federal, state, and local environmental regulators in terms of planning for site compliance.

3 THE TEAM AT MEGACORP

The customer's internal project team consists of:

- Environmental manager, Maureen Jones
- Corporate environmental lawyer, Paul Stanley
- Senior project manager, Tom Turbine
- Project engineer, Bill Preston

Maureen Jones has been with the Megacorp for only one year. Before coming to Megacorp, Ms. Jones had a high-ranking position with a large environmental consulting company, where she was employed for 12 years. Ms. Jones was able to convince Megacorp's upper level management that she knows how to effectively manage environmental consultants — to get the maximum amount of work for the minimum amount of cost. She is well respected within Megacorp for her take-charge style and technical knowledge. Ms. Jones has been challenged by upper level management to have remediation systems installed and operating at 15 company sites by the end of 1998. The legal department within Megacorp has an extraordinary amount of power. In fact, Ms. Jones has to answer to the supervisor of the legal department regarding any activities of the environmental staff.

Paul Stanley is a senior member of the legal group and is bucking for the supervisor's job. He is a sharp lawyer and is not afraid to stand his ground with regulators. Mr. Stanley strives to keep all his options open; he can be reluctant to commit to any objectives for the project — until he understands all associated risks and attached costs.

Tom Turbine is one of the most technically astute members of Megacorp's project team. Mr. Turbine has a background in mechanical engineering and had worked for five years in an engineering consulting firm that designed wastewater treatment systems. Since that position, he has been with Megacorp for 15 years. When Tom first came to Megacorp, he worked for five years designing industrial wastewater treatment systems for Megacorp's manufacturing plants. Ten years ago, he began his involvement with remediation projects.

Tom has numerous contacts throughout Megacorp. He is honest, intelligent, and is respected by both Maureen Jones and Paul Stanley. Tom has been given project management responsibility for a number of sites, including the one described later in this case study summary. He is, therefore, extremely busy and relies on a number of project engineers, such as Bill Preston, to handle the day-to-day communications with various environmental consultants contracted to work on the sites.

Bill Preston, the project engineer, has approximately three years of experience with Megacorp. He has a civil engineering degree specializing in environmental engineering. He is young and anxious to impress Maureen and Tom.

4 THE SALE

Assume that you work for an environmental consulting company known as Enviroserv and that Megacorp has been contacted by your company's director of marketing. In November of 1997, sales personnel made a presentation to Maureen, Tom, and Bill. The topics presented included new developments in remediation technology as well as strategies for least-cost closure. Megacorp seemed particularly impressed with your company's experience with the design and installation of high-vacuum dual-phase extraction technology. The attendees asked many questions regarding project management controls. The meeting progressed well, and the Megacorp personnel began to describe a site for which they planned to issue a request for proposal (RFP) in early 1998.

5 THE SITE

The site described by Megacorp personnel during the sales meeting is a former manufacturing plant (air compressors) located in a sparsely populated area of a midwestern state. The entire site covers approximately 35 acres, with the former manufacturing building covering approximately 3 acres. The site is bounded on all sides by open agricultural areas. The topography is relatively flat, with slightly elevated areas to the northwest of the former manufacturing building. There is a minor grade toward the Lower Takoma River, which is located approximately 1000 feet southeast of the site. The site map displayed during the sales meeting is provided as Figure 1. (All figures and tables appear at the end of this summary.)

6 PLANT OPERATIONAL HISTORY

The former manufacturing plant on the site was in operation from the early 1950s until 1987. This manufacturing building, along with approximately 20 acres of land, was sold to an auto parts company (Best Auto Parts, Inc.) in 1987. Best Auto Parts currently uses the former manufacturing building for warehousing purposes only. Megacorp retained liability for the cleanup of any subsurface environmental impacts that may have occurred at the facility while under Megacorp's control.

One environmental incident occurred while the site was still under Megacorp's ownership. In 1985, a leak occurred in a 12,000-gallon underground storage tank (UST) containing xylene. The UST was located directly east of the manufacturing building. The leak was verified during a partial excavation of the UST in January of that year. Additional investigations in 1985, performed by a small environmental consulting company that is no longer in business, verified significant impact to soil and groundwater in this area. In February 1986, the UST was removed and the excavation was filled with crushed limestone.

7 SITE CONDITIONS (NOVEMBER 1997)

All process machinery, associated equipment, and supplies owned by Megacorp were removed prior to the sale to Best Auto Parts. An electrical substation, located to the northeast side of the manufacturing building, has transformers filled with a dielectric fluid containing polychlorinated biphenyls (PCBs). There is no evidence that the transformers have leaked. Asphalt parking lots are located to the east and south side of the manufacturing building, and site vehicle access is limited by chain gates across the entrance drives.

The manufacturing building and the property to the north and west are currently owned by Best Auto Parts. Property located directly north and east of the parking area was retained by Megacorp. Megacorp's property is currently referred to as the former disposal area. Plant wastes (in the form of paints, miscellaneous fluids, parts, old tools, etc.) are known to have been deposited in four separate disposal pits, which had been excavated during Megacorp's plant operation.

Eighteen monitoring wells were installed at the site during prior environmental investigations, three of which are located in the former disposal area. These wells are the result of work by four environmental consultants who performed site investigations from January 1985 through December 1990. In addition, seven recovery wells were installed in the former xylene UST area by the first consultant involved at the site. The recovery wells are 1 inch in diameter and installed to a depth of approximately 9 feet. Because of poor installation practices and surface damage, none of the recovery wells are in functional condition. See Figures 1 and 2 for all well locations. Due to past performance issues, none of the previous environmental consultants are now involved at the site.

8 PREVIOUS ENVIRONMENTAL ACTIVITIES

A number of investigative and remedial activities occurred at the site beginning in early 1985 and proceeded through late 1990. This section briefly describes those activities.

The four environmental remediation companies that had worked for Megacorp are

- Envirofixx, Inc. (EFX)
- Terra Firma Corporation (TFC)
- Air, Land and Sea Associates (ALS)
- Eco-pure, Inc.

Envirofixx, Inc.

EFX was the first company called into the site by Megacorp. EFX performed the initial partial excavation of the tank and confirmed it was leaking. EFX was also responsible for the removal of the leaking tank. The details of the excavation were not well documented by EFX. Correspondence discovered in Megacorp's historical file indicated that the depth of excavation was 13.5 feet.

In the summer of 1985, EFX installed six monitoring wells (W-1, W-2, W-2A, W-3, W-3A, and W-4) in the vicinity of the former UST. Four of these wells were referred to as shallow wells (screened from 5 to 10 feet) and two of these wells were referred to as deep wells (screened from 12 to 18 feet, W-2A and W-3A). EFX was also responsible for installing the seven recovery wells which are of no use. EFX was dismissed after it recommended the installation of a "retaining box" in order to effect remediation of the site.

Terra Firma Corporation

In early 1986, TFC was hired to investigate the former disposal area. It installed ten trenches in this area with a backhoe. From these trenches, TFC collected 12 soil/waste samples. These samples were analyzed for PCBs only. The concern for PCBs at the site stems from the use of a cutting oil used in machining parts for the compressors. PCBs were a historical component of the cutting oil, used in order to make the oil nonflammable.

The highest concentration of PCBs found in the soil samples was reported at 68 ppm. TFC logged the presence of soil staining, drum debris, paint sludge, clinker, and other materials from the excavated trenches. Additionally, TFC claimed to have discovered the location of the historical disposal trenches used by the plant.

Upon completion of its investigation, TFC backfilled its trenches with the original excavated material. TFC did not develop quality maps regarding the location of its trenches; hence the locations can no longer be found. The analytical data obtained from the trench samples are still available. Megacorp promised the sales team that the analytical data would be contained in its RFP for the site.

Air, Land and Sea Associates

In July of 1986, ALS was hired by Megacorp to further define the extent of xylene migration. ALS installed six "shallow" monitoring wells (MW-1 through MW-6), located to the southwest, south, and southeast of the former UST. These six wells were screened from 5 to 10 feet. In addition, a seventh deep monitoring well (MW-7) was installed, 100 feet south of the former UST, which has a screened interval from 12 to 18 feet.

During the installation of monitoring wells MW-5 and MW-6, ALS collected two soil samples from each boring which it analyzed for ethyl benzene and xylene.

Eco-pure, Inc.

Finally, in 1990, Eco-pure, Inc. was hired to further investigate the area of the former UST and former disposal area, as well as sample all monitoring wells present on the site. Eco-pure installed six monitoring wells (MW-8 through MW-13), excavated 13 test pits in the former disposal area, and sampled all wells on site.

The groundwater samples from all wells were analyzed for Target Analyte List (TAL) volatiles and PCBs. In addition, samples from wells in the former disposal area were analyzed for TAL semivolatiles and metals. The groundwater samples, submitted for metals analysis, were not filtered in the field or the lab.

During the installation of monitoring wells MW-11, MW-12, and MW-13, soil samples were collected from the borings. These soil samples were analyzed for pesticides, PCBs, total metals, Toxic Characteristic Leachate Procedure (TCLP) metals, TAL volatiles, and semivolatiles. Information was not collected on background concentrations of metals in soil.

9 GEOLOGY AND HYDROGEOLOGY

The regional geology at Megacorp's site consists of glacial till, clay, sand, and gravel, from the surface to depths ranging from 100 to 140 feet. The drilling logs suggest that the geology directly beneath the site consists of sand and gravel lenses which are discontinuous laterally and are separated vertically by clay. Two water-bearing sand lenses have been reported in the vicinity of the former UST. The shallower of the two zones exists within a depth of 8 to 10 feet. The second exists within a depth of 12 to 14 feet. The zones are separated by a moist silty/clay layer.

A saturated sand lens has been reported in the former disposal area, at a depth of 12 to 14 feet. It is not known if this lens is in communication with either of the two lenses found in the former UST area. The regional groundwater flow is believed to be to the southeast toward the Lower Takoma River. However, in the vicinity of the former UST, groundwater flow appears to be radial away from the backfilled excavation. The groundwater elevation contours are presented in Figure 3.

10 ANALYTICAL CHARACTERIZATION

Megacorp agreed to supply bidders with soil and groundwater results from all sample locations along with the RFP. During the November 1997 meeting, Megacorp indicated that groundwater impact appeared to be limited to:

- The former UST area
- The wells monitoring the shallower of the two saturated lenses encountered within the former UST area

Megacorp reported that the groundwater has been found to contain primarily xylene; however, concentrations of ethyl benzene and PCBs have also been found. In addition, monitoring well MW-4, which is located directly south of the former UST and 200 feet from the site boundary, has shown elevated concentrations of xylene. Therefore, the extent of impact has not been fully determined.

11 REGULATORY COMPLIANCE ISSUES

The state where the site exists has recently developed a voluntary environmental cleanup program. This regulatory program is designed to encourage owners of impacted sites to begin to address their sites — without prior regulatory enforcement. Owners qualifying for this program, who conduct cleanups in a manner found acceptable to the state, can be awarded a certificate of closure from the governor, indicating that the case will not be reopened in the future. One of the fundamental qualifications for this program is that the site cannot currently be under an enforcement action whether issued by either the state or the U.S. EPA. If at all possible, Megacorp would like to qualify for this program. Unfortunately, Enviroserv, our consulting firm, has not had any experience with this program.

In early 1996, the state turned over copies of reports (developed by the previous consultants hired by Megacorp) to Region V of the EPA. Megacorp's lawyer, Paul Stanley, placed a call to the Region V director. He learned that the EPA is in the process of scoring the site for inclusion on the National Priorities List (NPL). Additionally, if the site is listed, the EPA director has indicated that EPA Region V would like to address this site in accordance with the new Superfund Accelerated Cleanup Model. Again, Enviroserv has not had any experience with sites where this new program was in place. Note that if the site ends up on the NPL list, it cannot be placed in the state's voluntary cleanup program.

Megacorp's internal regulatory council has determined that the xylene at the site is a listed hazardous waste (U239). Therefore, this project will involve elements of the Resource Conservation and Recovery Act. Furthermore, since PCBs have been found at the site, requirements under the Toxic Substances Control Act will also be a factor.

12 ENVIROSERV'S LOCAL RESOURCES

Enviroserv's nearest office is a three-hour drive from the site. This office has been involved primarily with petroleum UST projects. The office consists of two project managers, one engineer, one field geologist, and two field technicians.

13 THE MEGACORP REQUEST FOR PROPOSAL

On February 5, 1998, Enviroserv received an RFP for the Megacorp former compressor manufacturing site. All proposals are due in Megacorp's corporate office (in Chicago) on February 20, 1998.

13.1 Stated Request for Proposal Objectives

In the RFP, Megacorp stated its desire to contract a qualified environmental consultant for the development and execution of a work plan to perform a final investigation of the compressor site. The intent of this work plan is to fully determine the nature and extent of impact. The scope of work presented in the successful consultant's proposal will mirror the work plan to be submitted to the appropriate regulatory agency The work plan is to include special focus on the former xylene UST area and the former disposal area.

Since the successful bidder must fully develop the work plan in order to provide a cost estimate and schedule for its execution, the successful bidder will merely be required to reformat the scope of work contained within its proposal for submission to the regulatory agency. The RFP has emphasized that the job is being bid in total. In other words, one company will be awarded the development of the work plan and its execution. The contract will be awarded on a firm fixed price basis.

The RFP has also requested that the work plan include pilot testing and design of an appropriate technology to address groundwater impacts in the former xylene UST area. Therefore, based on the data provided in the RFP, the bidding contractors will have to select the appropriate technology and develop a schedule and cost estimate for pilot testing and design. A cost estimate for installation is not requested at this time. Finally, the RFP requested the development of a cost estimate be provided for a site health and safety plan to correspond to the investigative work plan.

A two-week response time has been requested for proposal submission. At the present time, your company has not negotiated a standard contract form with Megacorp.

13.2 Information Provided in the Request for Proposal

The following information was included in the RFP:

- Maps of the site (Figures 1 and 2)
- Groundwater gradient map (Figure 3)
- Groundwater compound distribution map (Figure 4)
- Tables of analytical results from:
  - Soil/waste samples collected by TFC from trenches installed in the former disposal area (Table 1)
  - Soil samples collected from test pits excavated by Eco-pure (Table 1)
  - Soil samples collected by ALS from MW-5 and MW-6 borings (Table 1)

    □ The last round of groundwater sampling from all wells on site (1990) (Tables 2 and 3)

It should be noted that semivolatile compounds were not discovered in the groundwater during the 1990 sampling event.

13.3   Other Information

Through a series of phone calls and meetings with Megacorp, both prior to and after receipt of the RFP, the following information has been gathered:

- Consistent with its proactive approach, the Megacorp project team has indicated that it would begin executing the work plan prior to regulatory approval.
- Megacorp is expecting a recommendation from the bidding contractors regarding the need for a quality assurance project plan in conjunction with the investigative work plan.
- One of the major criteria used by Megacorp in evaluating the proposals is demonstration of a comprehensive logical approach and project management skills.
- Tom Turbine, Megacorp's senior project manager, is impressed with Enviroserv's experience with high-vacuum dual-phase extraction technology and would like to see it pilot tested at the site. In fact, the sales representative has indicated that this process may be the primary reason Enviroserv was offered the chance to bid on this project.
- Maureen Jones, Megacorp's manager of environmental affairs, would like to accelerate the project as much as possible and is interested in installing a system to address the former xylene area before the end of the year.
- Maureen Jones is also interested in using a new surface geophysical tool at the site known as a gigometer to better identify buried debris in the former disposal area. Discussions with internal senior technical personal indicate that the gigometer has generated much industry interest, but thus far applications of this technique have been less than satisfactory.
- It is still unknown whether the site will be handled by EPA Region V, which will want to address the site in accordance with the new Superfund Accelerated Cleanup Model, or if the state will take the lead.
- Paul Stanley, Megacorp's lawyer, has contacted EPA Region IV and, based on his discussion, is convinced the EPA will issue a consent order

for the site before year-end. Paul would rather deal with EPA Region V based on his contacts within this group. In addition, Paul believes that the new Superfund process best suits Megacorp's proactive approach.

- Paul Stanley does not want to agree to any cleanup goals for the remediation system that may be installed at the site.
- It was discovered late in late March 1996 that the state sampled the sediments of Little Takoma River near the site and found significant PCB impacts. Historically, the state has been most concerned about PCBs at the site and less concerned about the xylene impact. In addition, the state reported to Paul Stanley that it has information which indicates that the electric substation had some leaking PCB transformers in the past.
- The management at Best Auto Parts has expressed concerns over disruption during investigative and remediation activities and has requested to be informed of all planned activities.
- A group of workers who were laid off as a result of Megacorp's decision to sell the facility is said to be preparing a lawsuit against Megacorp — due to chemical exposure while working at the site.
- The state's air permit requirements allow the discharge of 2 pounds per hour and a maximum of 20 pounds a day of volatile organic carbon compounds such as xylene.
- An access point to a sanitary sewer is available on-site. The local publicly owned treatment works does not have authority from the state to set its own pretreatment standards. Therefore, all permits to discharge to the sanitary sewer must be submitted to the state. Approval can require three months.

14 THE COMPETITION

Megacorp has chosen to send the RFP to three companies, the others being Eco-pure and Soilrem, Inc. It is curious that Eco-pure was invited back to bid on the site. It was originally dismissed for not maintaining sufficient documentation regarding the location of the test pits it excavated in the former disposal area.

Eco-pure is very familiar with the site and the quality of its work has improved over the last several years. Note that Eco-pure is better known for investigation than technology implementation. Soilrem is a new player in the market and there is some evidence that the principals in this company are former associates of Maureen Jones.

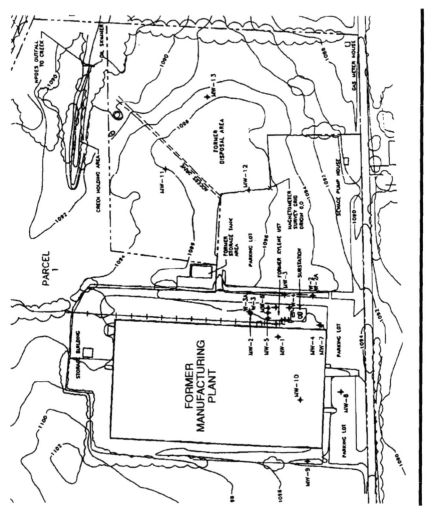

FIGURE 1 Site plan.

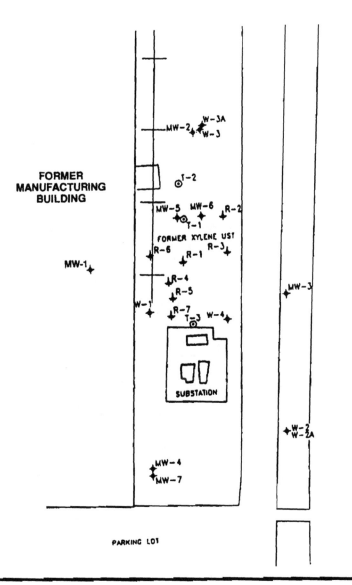

FIGURE 2 Xylene area.

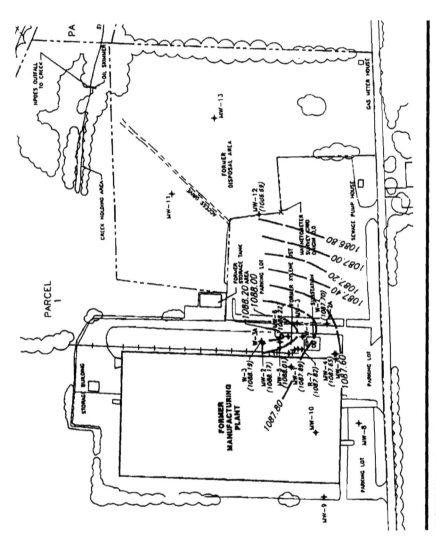

FIGURE 3 Groundwater elevation contours.

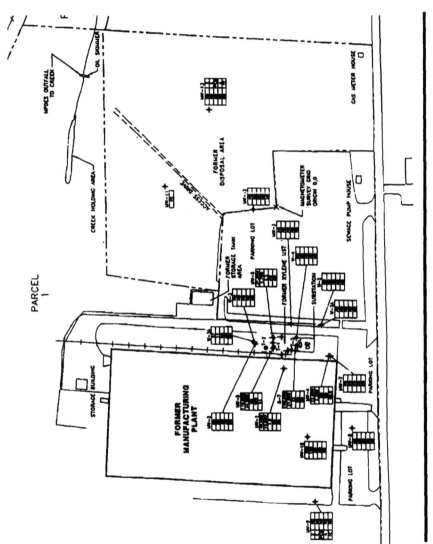

FIGURE 4 Compound distribution.

TABLE 1 Soil Analytical Results

Location (numerical) Date (chronological)	TR-1-6' 3-14-86	TP-1-2/4' 7-24-90	TP-2-0/2' 7-24-90	TP-2-4/6' 9-21-90	TR-3-1.5' 3-14-86	TR-3-4' 3-14-86	TP-3-0/2' 7-24-90	TP-4-2/4' 7-24-90	TP-4-4/6' 9-21-90
Compound									
Pesticides (ppm)	ND	ND	ND	ND			ND	ND	ND
PCBs (ppm)									
PCB 1242	ND	ND	ND	ND	4.5	ND	ND	ND	ND
PCB 1248	ND	ND	ND	ND	ND	ND	ND	ND	ND
PCB 1254	0.32	ND	8.5	ND	4.5	ND	ND	ND	ND
Total metals (ppm)									
Aluminum		1,900	15,000	26,000			12,000	12,000	23,000
Antimony		ND	ND	ND			ND	ND	8
Arsenic		25	19	41			17	9	13
Barium		130	130	140			100	120	89
Beryllium		0.9	0.3	0.8			0.4	0.4	0.8
Cadmium		ND	ND	ND			ND	ND	ND
Calcium		5,900	66,900	26,000			45,700	53,400	44,000
Chromium		20	120	27			15	13	27
Cobalt		15	11	11			13	14	9
Copper		26	150	24			17	26	24
Iron		32,100	16,100	28,000			18,600	22,100	25,000
Lead		12	83	8			10	7	9
Magnesium		4,400	32,000	10,000			21,000	26,800	27,000
Manganese		640	330	340			1,100	520	280
Mercury		0.03	0.07	ND			ND	ND	ND
Nickel		30	16	27			17	29	24
Potassium		1,100	510	5,100			1,100	1,100	5,000
Silver		ND	ND	ND			ND	ND	ND
Sodium		590	510				440	460	
Vanadium		34	21	47			27	25	47
Zinc		84	220	91			57	71	36

	TR-5-2'	TR-5-3'	TP-5-2/4'	TP-5-4/6'	TP-6-2/4'	TP-6-4/6'	TR-7-4'	TR-7-4.5'	TR-7-5'
TCLP metals (ppm)									
Arsenic			ND				ND	ND	ND
Barium			0.8				ND	ND	0.6
Cadmium			ND				ND	ND	ND
Chromium			ND				ND	ND	ND
Selenium			ND				ND	ND	ND
Silver			ND				ND	ND	ND
Volatile compounds (ppm)									
Ethyl benzene		ND		ND			ND	ND	ND
Xylenes (total)		ND		ND			ND	ND	ND
Semivolatiles (ppm									
Benzyl alcohol		ND		ND			ND	ND	ND
Bis(2-ethylhexyl)phthalate		ND		ND			ND	ND	ND
Isophorone		ND		ND			ND	ND	ND
2-Methylnaphthalene		ND		ND			ND	ND	ND
4-Methylphenol		ND		ND			ND	ND	ND
Naphthalene		0.47		ND			ND	ND	ND
Phenanthrene		ND		ND			ND	ND	ND
Location (numerical)	*TR-5-2'*	*TR-5-3'*	*TP-5-2/4'*	*TP-5-4/6'*	*TP-6-2/4'*	*TP-6-4/6'*	*TR-7-4'*	*TR-7-4.5'*	*TR-7-5'*
Date (chronological)	*3-14-86*	*3-14-86*	*7-24-90*	*7-24-90*	*7-25-90*	*9-21-90*	*3-14-86*	*3-14-86*	*3-14-86*
Pesticides (ppm)			ND	ND	ND	ND			
PCBs (ppm)									
PCB 1242	0.58	ND	ND	ND	ND	ND	ND	ND	ND
PCB 1248	ND	ND	ND	ND	ND	ND	ND	3.2	36
PCB 1254	0.66	ND	ND	ND	ND	ND	0.3	ND	ND
Total metals (ppm)									
Aluminum			7,800	5,800	17,600	2,200			
Antimony			ND	67	ND	20			
Arsenic			10	ND	16	5			
Barium			71	38	110	23			
Beryllium			0.3	ND	0.7	ND			

TABLE 1 Soil Analytical Results (continued)

Location (numerical) Date (chronological)	TR-5-2' 3-14-86	TR-5-3' 3-14-86	TP-5-2/4' 7-24-90	TP-5-4/6' 7-24-90	TP-6-2/4' 7-25-90	TP-6-4/6' 9-21-90	TR-7-4' 3-14-86	TR-7-4.5' 3-14-86	TR-7-5' 3-14-86
Cadmium			ND	ND	ND	ND			
Calcium			95,700	14,700	5,900	77,000			
Chromium			8	10	19	3			
Cobalt			13	ND	14	4			
Copper			19	31	22	5			
Iron			16,900	10,000	23,000	6,100			
Lead			5	9	13	3			
Magnesium			35,000	66,000	4,800	27,000			
Manganese			380	160	720	184			
Mercury			ND	ND	ND	ND			
Nickel			23	6	31	9			
Potassium			770	1,700	390	210			
Silver			ND	ND	ND	12			
Sodium			380		280				
Vanadium			18	14	29	6			
Zinc			60	79	81	27			
TCLP metals (ppm)									
Arsenic			ND		ND				
Barium			ND		0.9				
Cadmium			ND		ND				
Chromium			ND		ND				
Selenium			ND		ND				
Silver			0.1		ND				
Volatile compounds (ppm)									
Ethyl benzene			ND	ND	ND	ND			
Xylenes (total)			ND	ND	ND	ND			
Semivolatiles (ppm)									
Benzyl alcohol			ND	ND	ND	ND			

	TR-7-7* 3-14-86	TP-7-4/6' 7-25-90	TP-7-10/12' 9-21-90	TP-8-4/6' 7-25-90	TR-9-4.5' 3-14-86	TR-9-10* 3-14-86	TP-9-4/6' 7-26-90	TP-9-10/12' 9-21-90
Bis(2-ethylhexyl)phthalate				ND		0.28		ND
Isophorone				ND		ND		ND
2-Methylnaphthalene				ND		ND		ND
4-Methylphenol				ND		ND		ND
Naphthalene				ND		ND		ND
Phenanthrene				ND		ND		ND
Location (numerical) / *Date (chronological)*								
Pesticides (ppm)				ND				ND
PCBs (ppm)								
PCB 1242	0.023	ND	ND	ND	ND	0.059	ND	ND
PCB 1248	ND	ND	1.6	ND	6.7	ND	ND	40.6
PCB 1254	ND	11.1	ND	4	ND	ND	ND	ND
Total metals (ppm)								
Aluminum		9,800	1,200	15,000			17,100	19,000
Antimony		ND	17	ND			ND	37
Arsenic		15	13	19			27	20
Barium		200	99	120			110	1,800
Beryllium		0.4	0.5	0.5			0.7	0.9
Cadmium		2	ND	ND			ND	5
Calcium		100,000	99,000	36,700			17,000	15,000
Chromium		13	12	18			19	510
Cobalt		16	6	13			13	18
Copper		160	27	27			55	47,000
Iron		20,800	15,000	24,000			27,700	145,000
Lead		31	6	10			15	440
Magnesium		29,000	40,000	13,000			8,900	7,700
Manganese		320	110	420			540	1,300
Mercury		0.8	ND	0.07			ND	0.2
Nickel		29	13	26			31	67
Potassium		1,200	3,400	1,100			1,700	1,300

TABLE 1 Soil Analytical Results (continued)

Location (numerical)	TR-7-7*	TP-7-4/6'	TP-7-10/12'	TP-8-4/6'	TR-9-4.5'	TR-9-10**	TP-9-4/6'	TP-9-10/12'
Date (chronological)	3-14-86	7-25-90	9-21-90	7-25-90	3-14-86	3-14-86	7-26-90	9-21-90
Silver	1		2	ND			ND	5
Sodium		430		440			790	
Vanadium		21	25	27			34	30
Zinc		40	ND	88			100	20,000
TCLP metals (ppm)								
Arsenic		ND		ND			ND	
Barium		2.2		1.4			0.8	
Cadmium		ND		ND			ND	
Chromium		ND		ND			ND	
Selenium		ND		ND			ND	
Silver		ND		ND			ND	
Volatile compounds (ppm)								
Ethyl benzene		1.01	ND	ND			ND	4.5
Xylenes (total)		23.1	ND	ND			9.26	34.1
Semivolatiles (ppm)								
Benzyl alcohol		ND	ND	ND			ND	1.69
Bis(2-ethylhexyl)phthalate		ND	ND	ND			6.2	1.25
Isophorone		ND	ND	ND			ND	ND
2-Methylnaphthalene		0.84	ND	ND			1	ND
4-Methylphenol		ND	ND	ND			ND	20.46
Naphthalene		4.2	ND	ND			0.99	0.65
Phenanthrene		0.61	ND	ND			0.42	ND

Location (numerical)	TR-10-5.5'	TP-10-8/10'	TP-10-4/6'	TR-11-5'	TP-11-4/6'	TP-11-8/10'	TP-11-10/12'	TR-12-6'
Date (chronological)	3-14-86	7-26-90	9-21-90	3-14-86	9-21-90	9-21-90	9-21-90	3-14-86
Pesticides (ppm)		ND	ND	ND	ND	ND	ND	ND

	1	2	3	4	5	6	7	8
PCBs (ppm)								
PCB 1242	0.35	ND	ND	ND	ND	ND	ND	ND
PCB 1248	ND	ND	6.9	19	9.1	4.6	7	68
PCB 1254	1.9	ND	ND	ND	ND	ND	ND	ND
Total metals (ppm)								
Aluminum	5,800		27,000		2,200	9,900	3,200	
Antimony	ND		26		15	12	15	
Arsenic	10		32		18	30	5	
Barium	48		1,600		2,500	86	57	
Beryllium	0.3		0.9		0.9	ND	0.3	
Cadmium	ND		4		3	ND	ND	
Calcium	95,000		35,000		36,000	100,000	110,000	
Chromium	6		170		220	19	6	
Cobalt	9		28		27	7	4	
Copper	21		3,700		3,900	110	26	
Iron	13,100		85,000		61,000	19,000	9,700	
Lead	4		690		980	23	10	
Magnesium	30,000		27,000		34,000	35,000	27,000	
Manganese	230		880		960	240	290	
Mercury	ND		0.2		ND	0.04	ND	
Nickel	16		83		51	17	11	
Potassium	1,000		1,700		1,300	2,900	750	
Silver	ND		3		2	ND	0.3	
Sodium	330							
Vanadium	15		41		33	10	10	
Zinc	48		220		220	130	59	
TCLP metals (ppm)								
Arsenic	ND							
Barium	ND							
Cadmium	ND							
Chromium	ND							
Selenium	ND							
Silver	ND							

TABLE 1 Soil Analytical Results (continued)

Location (numerical) Date (chronological)	TR-10-5.5' 3-14-86	TP-10-8/10' 7-26-90	TP-10-4/6' 9-21-90	TR-11-5' 3-14-86	TP-11-4/6' 9-21-90	TP-11-8/10' 9-21-90	TP-11-10/12' 9-21-90	TR-12-6' 3-14-86
Volatile compounds (ppm)								
Ethyl benzene		ND	2.92		5.04	12.1	ND	
Xylenes (total)		ND	19.4		23.8	45.1	ND	
Semivolatiles (ppm)								
Benzyl alcohol		ND	ND		ND	ND	ND	
Bis(2-ethylhexyl)phthalate		ND	0.56		0.32	0.63	ND	
Isophorone		ND	ND		ND	0.76	ND	
2-Methylnaphthalene		ND	4.29		0.8	3.39	ND	
4-Methylphenol		ND	ND		ND	ND	ND	
Naphthalene		ND	14.29		2.25	6.78	ND	
Phenanthrene		ND	ND		ND	0.49	ND	

Location (numerical) Date (chronological)	TP-12-4/6' 9-21-90	TP-12-8/10' 9-21-90	TR-13-5.5' 3-14-86	TP-13-5.5' 3-14-86	TP-13-4/6' 7-30-90	TP-13-6/8' 9-21-90	TR-14-4' 3-14-86	MW-5-5/6.5' 7-10-86	MW-5-7.5/9' 7-10-86
Pesticides (ppm)	ND	ND			ND	ND	ND		
PCBs (ppm)									
PCB 1242	ND	ND	ND		ND	ND	3.8		
PCB 1248	ND	ND	3.2		ND	ND	ND		
PCB 1254	ND	ND	ND		26.9	ND	ND		
Total metals (ppm)									
Aluminum	5,500	18,000			42,600	15,000			
Antimony	11	ND			38	ND			
Arsenic	22	ND			37	36			
Barium	42	83			2,500	78			
Beryllium	0.4	0.7			0.6	0.5			
Cadmium	0.6	ND			6	ND			
Calcium	41,000	69,000			260,000	67,000			

	MW-6-5/6.5'	MW-6-7.5/9'	MW-11-10/12'	MW-12-2/4'	MW-12-6/8'	MW-13-4/6'	MW-13-14/16'
Chromium	34	17	69			15	
Cobalt	6	6	34			8	
Copper	36	20	2,800			20	
Iron	26,800	21,000	27,900			18,000	
Lead	8	5	380			5	
Magnesium	22,000	33,000	37,000			32,000	
Manganese	210	20	590			400	
Mercury	0.08	20	1.1			ND	
Nickel	22	20	72			20	
Potassium	5,100	4,800	580			4,100	
Silver	ND	ND	3			ND	
Sodium			1,800				
Vanadium	11	37	32			34	
Zinc	150	ND	3,100			67	
TCLP metals (ppm)							
Arsenic			0.05				
Barium			0.78				
Cadmium			0.02				
Chromium			0.03				
Selenium			ND				
Silver			ND				
Volatile compounds (ppm)							
Ethyl benzene	ND	ND	2.06	0.028	3,050		59
Xylenes (total)	ND	ND	2.36	0.7			
Semivolatiles (ppm)							
Benzyl alcohol	ND	ND	ND				
Bis(2-ethylhexyl)phthalate	0.46	1.22	3.6				
Isophorone	ND	ND	ND				
2-Methylnaphthalene	ND	ND	7.9				
4-Methylphenol	ND	ND	ND				
Naphthalene	0.99	ND	28				
Phenanthrene	0.42	ND	0.34				
Location (numerical)	MW-6-5/6.5'	MW-6-7.5/9'	MW-11-10/12'	MW-12-2/4'	MW-12-6/8'	MW-13-4/6'	MW-13-14/16'

TABLE 1 Soil Analytical Results (continued)

Date (chronological)	7-10-86	7-10-86	7-26-90	7-27-90	7-27-90	7-27-90	7-27-90
Pesticides (ppm)			ND	ND	ND	ND	ND
PCBs (ppm)							
PCB 1242			ND	ND	ND	ND	ND
PCB 1248			ND	ND	ND	ND	ND
PCB 1254			1.9	ND	ND	ND	ND
Total metals (ppm)							
Aluminum			5,500	7,600	4,000	10,800	2,900
Antimony			ND	ND	ND	ND	ND
Arsenic			8	13	4	20	ND
Barium			36	53	33	82	42
Beryllium			ND	ND	ND	0.4	ND
Cadmium			ND	ND	ND	ND	ND
Calcium			130,000	110,000	136,000	73,300	113,000
Chromium			7	9	5	12	2
Cobalt			10	11	10	11	8
Copper			14	15	12	15	11
Iron			12,000	15,000	10,000	18,000	2,600
Lead			4	5	7	7	6
Magnesium			36,000	33,000	43,000	29,500	48,000
Manganese			240	240	230	195	110
Mercury			ND	ND	ND	ND	ND
Nickel			18	25	11	23	13
Potassium			1,100	1,400	540	1,400	990
Silver			ND	ND	ND	ND	ND
Sodium			340	370	300	410	290
Vanadium			15	20	11	23	6
Zinc			46	76	29	57	32
TCLP metals (ppm)							
Arsenic			ND	0.03	ND	ND	ND

Location (numerical) / Date (chron)	MW-6-5/6.5' 7-10-86	MW-6-7.5/9' 7-10-86	MW-11-10/12 7-26-90	MW-12-2/4' 7-27-90	MW-12-6/8' 7-27-90	MW-13-4/6' 7-27-90	MW-13-14/16' 7-27-90	Total number of samples	Average concentration detected at site	Minimum concentration detected at site	Maximum concentration detected at site	Location of maximum
Barium			ND	ND	ND	ND	0.8					
Cadmium			ND	ND	ND	ND	ND					
Chromium			ND	ND	ND	ND	ND					
Selenium			ND	ND	ND	ND	0.03					
Silver			ND	0.1	ND	ND	ND					
Volatile compounds (ppm)												
Ethyl benzene			ND	ND	ND	ND	ND					
Xylenes (total)	862	516	ND	ND	ND	ND	ND					
Semivolatiles (ppm)												
Benzyl alcohol			ND	ND	ND	ND	ND					
Bis(2-ethylhexyl)phthalate			ND	ND	ND	ND	ND					
Isophorone			ND	ND	ND	ND	ND					
2-Methylnaphthalene			ND	ND	ND	ND	ND					
4-Methylphenol			ND	ND	ND	ND	ND					
Naphthalene			ND	ND	ND	ND	ND					
Phenanthrene			ND	ND	ND	ND	ND					
Pesticides (ppm)								30	ND	ND	ND	
PCBs (ppm)												
PCB 1242								45	0.21	ND	4.5	TR-3-1.5'
PCB 1248								45	4.58	ND	68	TR-12-6'
PCB 1254								45	1.34	ND	26.9	TR-13-4/6'
Total metals (ppm)												

TABLE 1 Soil Analytical Results (continued)

	Total number of samples	Average concentration detected at site	Minimum concentration detected at site	Maximum concentration detected at site	Location of of maximum
Aluminum	29	11,910.34	1,200	42,600	TP-13-4/6'
Antimony	29	9.17	ND	67	TP-5-4/6'
Arsenic	29	16.69	ND	41	TP-2-4/6'
Barium	29	362.76	23	2,500	TP-11-4/6'
Beryllium	29	0.43	ND	0.9	TP-11-4/6'
Cadmium	29	0.71	ND	6	TP-13-4/6'
Calcium	29	71,662.07	5,900	260,000	TP-13-4/6'
Chromium	29	49.14	2	510	TP-9-10/12'
Cobalt	29	12.03	ND	34	TP-13-4/6'
Copper	29	2,010.79	5	47,000	TP-9-10/12'
Iron	29	26,527.59	2,600	145,000	TP-9-10/12'
Lead	29	96.38	3	980	TP-11-4/6'
Magnesium	29	28,244.83	4,400	66,000	TP-5-4/6'
Manganese	29	419.97	20	1,300	TP-9-10/12'
Mercury	29	0.78	ND	20	TP-12-8/10'
Nickel	29	191.72	6	4,800	TP-12-8/10'
Potassium	29	1,673.79	ND	5,100	TP-12-4/6'
Silver	29	0.98	ND	12	TP-6-4/6'
Sodium	16	510.00	280	1,800	TP-13-4/6'
Vanadium	29	24.24	6	47	TP-2-4/6'
Zinc	29	871.31	ND	20,000	TP-9-10/12'
TCLP metals (ppm)					
Arsenic	16	0.01	ND	0.05	TP-13-4/6'

Barium	16	0.52	ND	2.2	TP-7-4/6'
Cadmium	16	ND	ND	0.02	TP-13-4/6'
Chromium	16	ND	ND	0.03	TP-13-4/6'
Selenium	16	ND	ND	0.03	MW-13-14/16'
Silver	16	0.01	ND	0.1	MW-12-2/4'
Volatile compounds (ppm)					
Ethyl benzene	29	0.95	ND	12.1	TP-11-8/10.5'
Xylenes (total)	33	140.75	ND	3,050	MW-5-5/6.5'
Semivolatiles (ppm)					
Benzyl alcohol	29	0.06	ND	1.69	TP-9-10/12'
Bis(2-ethylhexyl)phthalate	29	0.50	ND	6.2	TP-9-4/6'
Isophorone	29	0.03	ND	0.76	TP-11-8/10.5'
2-Methylnaphthalene	29	0.63	ND	7.9	TP-13-4/6'
4-Methylphenol	29	0.71	ND	20.46	TP-9-10/12'
Naphthalene	29	2.03	ND	28	TP-13-4/6'
Phenanthrene	29	0.07	ND	0.61	TP-7-4/6'

Note: ND = not detected, blank = not analyzed, TR = trench, TP = test pit, * = water from trench.

TABLE 2 TAL VOCs and PCBs in Groundwater: September 1990 Sampling Event (Concentration in µg/l)

Analyte	MW-1	MW-2	MW-3	MW-4	MW-5	MW-6	MW-7	MW-8	MW-9
Methylene chloride					54				
Acetone					400				
1,1-Dichloroethane				15					
1,2-Dichloroethane				19					
1,2-Dichloroethene				6.9					
Toluene	20			6.7	110				
Benzene	26			58					
Ethyl benzene	16,000			27,000	34,000				
Xylenes	40,000			160,000	110,000				
Arochlor 1260	4.8					4.5			

Analyte	MW-10	MW-12	MW-13	R-7	W-2	W-2A	W-3	W-3A	W-4
Methylene chloride									
Acetone									
1,1-Dichloroethane									
1,2-Dichloroethane									
1,2-Dichloroethene									
Toluene									
Benzene									
Ethyl benzene	190			27,000				14	
Xylenes	700			160,000				87	
PCBs Arochlor 1260				42.0					

Note: Blank indicates analyte below reportable limit.

TABLE 3   TAL Total Metals in Groundwater: September 1990 Sampling Event (Concentration in μg/l)

Analyte	MW-1	MW-4	MW-12	MW-12	MW-13
Arsenic			62	170	26
Barium			650	450	750
Cadmium			1	1	
Calcium (ppm)					460
Chromium			58	180	24
Lead			36	100	80
Mercury			0.5	0.4	
Selenium				12	
Silver					
Sodium (ppm)			8.4	22	
Zinc	473	257			
Reactivity cyanide	20	30			

Note: Table includes only wells which had reportable concentrations of one or more metals. Blank indicates analyte below reportable limit.

APPENDIX B

Project Planning Forms

Bid - No Bid Evaluation Form

Page 1 of 2

A. Milestones

Date RFP Issued:	Date RFP Received:	Bid Decision Date:	Date Proposal Due:

B. Proposal Information

Lead Person: Proposal No.:   Customer: Customer Contact:

Telephone No.:   Location: ProjectName:

Project Description:

Competitors:

Special Considerations: Unique Company Qualifications:

Last Visited By: Visit Date:

C. Bid - No Bid Decision

Bid Factors	Matrix Weighted Decision Criteria			Your Company	Competitor A	Competitor B	
	Negative 0-3	Neutral 4-6	Positive 7-10				
1. Previous experience with customer?	Poor	Average	Good				
2. Previous experience with similar projects?	Poor	Average	Good				
3. Opportunity was pre-sild by marketing?	No	Average	Yes				
4. Alignment with your company's business plan?	Poor	Fair	Good				
5. Pace competition/number of competitors?	Strong	Medium	Weak				
6. Are we bidding against an incumbent?	Yes	Maybe	No				
7. Conflict of interest?	Yes	Possibly	No				
8. Availability of required operational resources?	Poor	Average	Good				
9. Availability of technical expertise?	Poor	Average	Good				
10. Customer's financial condition, stability, and reputation?	Weak	Medium	Strong				
11. Anticipated follow - on work?	No	Possibly	Yes				
12. Company office location?	Unfavorable	Neutral	Favorable				
13. Rate schedule/profitability?	Unfavorable	Neutral	Favorable				
14. Potential for new technology or new markets?	Poor	Average	Good				
15. Does customer have funds approved?	No	Don't Know	Yes				
16. Does the project require regulatory approval?	Yes	Don't Know	No				
17. If yes, do they have approval?	No	Don't Know	Yes				
				Total Estimated Score			
				Maximum Score Available (# of factors X 10)			

Comments:

Prepared by:	Reviewed by:	Date	Preliminary Decision:
Prepared by:	Reviewed by:	Date:	Bid No Bid

D. Intermediate (Contract) Bid - No Bid Decision

Key Contract Evaluation Criteria   Circle Appropriate Rating

18. Terms and Conditions	Negative Unfavorable	Neutral Average
19. Insurance and Bonding Requirements	Not Acceptable	Average
20. Significant Contract Negotiation Required	Yes, Not Allowed	Some
21. Level of Potential Risk (including 3rd party)	High	Average

Positive
Favorable
Acceptable
No or Not Required
Low

Comments:

Contract Type (circle one):   Firm Fixed Price   Time & Materials   Cost Plus Fixed Fee

Intermediate Decision (circle one):   Bid   No Bid:

Contract Review: Date:

Bid - No Bid Evaluation Form

Page 2 of 2

E. Estimated Preparation Costs

Proposal Team	Name	Billing Category	Hours	Rate	Cost	TOTAL:
Proposal Manager						
Sales/Marketing						
Contracts						
Technical I						
Technical II						
Technical III						
Technical IV						
Cost/Accounting						
Document Production						
Administrative						
Other Costs (i.e., Travel)						

F. Estimated Project Revenues and Profit

	Revenue (X)	Margin (%)	(=) Profit
Your Company's Professional Services			
Subcontractors/Vendors			

G. Estimated Profit To Proposal Cost Ratio

Estimated Profit (X) Probability of Award
—————————————————————————— = $\dfrac{\$}{\$}$ =
Total Proposal Cost

H. Final Bid - No Bid Decision

Final Decision (circle one): BID   NO BID

Comments:

Technical Review: Date:
Operations Review: Date:
Proposal Manager Review: Date:
Operations Review: Date:

RFP Summary Form	
Background Information	
Operational History	
Known	*Unknown*

RFP Summary Form
Background Information
Current Site Conditions

Known	*Unknown*

RFP Summary Form	
Background Information	
Previous Environmental Activities (if any)	
Known	*Unknown*

Page 3 of 7

RFP Summary Form
Physical Characterization
Regional Geology

Known	*Unknown*

Site Geology

Known	*Unknown*

Page 4 of 7

RFP Summary Form	
Analytical Characterization	
Soil Analytical Results	
Known	*Unknown*
Groundwater Analytical Results	
Known	*Unknown*
Sediment Analytical Results	
Known	*Unknown*

Page 5 of 7

RFP Summary Form
Regulatory Issues

Known	*Unknown*

Page 6 of 7

RFP Summary Form
Regulatory Issues

Known	*Unknown*

Stakeholder Management Summary Form			
Stakeholder Name	**Specific Stake**	**Strengths**	**Weaknesses**

Stakeholder Management Summary Form		
Stakeholder Name	Expected Strategy or Behavior	Stakeholder Management Response Strategies

Page 2 of 2

APPENDIX C

Pre-Bid Site Investigation Checklist

The following checklist will serve as a guide in evaluating the variables associated with a project. In addition to completing this checklist, the investigators should make a videotape of the project site, access roads, and any other features of the area that may influence final evaluation of the project. Ideally, this tape should be narrated as shot.

Individuals participating in the site investigation should become very familiar with the bid documents, especially the actual contract and specifications, before beginning their investigation so they can properly relate area conditions to contractual requirements. In taking the actions or responding to the questions in this checklist, the investigators must continually seek to identify and catalog all items that have a cost and time implication for the contractor.

1. Geographic data
    A. Obtain the following maps and annotate them with responses to questions appearing later in the checklist:
       (1) State maps to show overall highway, railroad, and water networks
       (2) Local maps to show details of nearby communities and local road networks
       (3) Topographic (contour) map

From *Skills and Knowledge of Cost Engineering,* 3rd ed., 1992. Reprinted with the permission of AACE International, 209 Prairie Ave., Suite 100, Morgantown, WV USA. Phone: 800-858-COST/304-296-3444. Fax: 304-291-5728. E-mail: 74757.2636@compuserve.com. Internet: http://www.aacei.org or www.cost.org.

    B. Provide the population of each nearby community.

    C. Trace and describe the existing drainage system for the area. Will drainage be an unusual problem?

    D. What is the site altitude?

2. Geological and subsurface data

    A. Obtain a geological map of the area.

    B. What are the subsurface conditions at the project site — soils strata and characteristics, water table? Determine if any part of the site is a fill area. If so, show its location and describe the type of fill.

    C. Will dewatering be required?

    D. For excavations, what methods or equipment will be required for soil removal?

    E. Will shoring be needed for ditches and other excavations?

    F. Will the natural ground support construction equipment, both rubber-tired and tracked?

    G. Identify the drainage characteristics of the topsoil following precipitation. How long does it take for the surface to dry out? Will topsoils be particularly dusty when dry if vegetation is removed?

    H. Will a temporary construction drainage system be needed?

    I. Obtain a plot of all known underground utility and drainage systems or other underground structures or obstacles.

3. Meteorological data

    A. Describe temperature and humidity ranges by month. Include record low and high temperature for each month; if wind chill or heat index data are available, include them.

    B. Describe rainfall and snow averages by month. Determine the average number of precipitation days for each month.

    C. What is the susceptibility of the area to and frequency of extreme storm activity — thunderstorms, snow storms, tornadoes, and hurricanes?

    D. What is the frost depth?

4. Site preparation

    A. Can excess cut material be sold? If so, to whom and at what price? If not, where are disposal areas located? What are the disposal charges involved?

    B. If fill material is required, where are its sources located? What is its cost?

    C. What overhead lines or other structures will limit operations or prove to be a safety hazard in clearing the site?

    D. Will erosion control structures be required?

    E. What historical structures, graveyards, plantings, or other features must be moved or protected during site preparation or while construction is underway?

5. Access and parking

    A. Describe existing access roads in terms of lanes, surface type and condition, capacity limitations, restrictive features, etc. What must be done to upgrade them for contractor use? Will the contractor have any responsibility for maintaining these roads?

    B. What new access roads must be developed and who will develop them? Are existing easements or rights-of-way available for such roads? If not, what must be done to obtain them?

    C. Is adequate parking space available on the site? If not, where must workers park and how do they get to the site? What parking space development is still required and who will do it? Is the parking area fenced or must fencing be added?

6. Unloading and storage

    A. Can material be received and stockpiled on the site before construction begins? If so, what security will be required and who will provide it? Will the owner provide labor to receive these materials? Will this create union jurisdictional problems?

    B. What on-site areas are available for development by the contractor into laydown or other storage/prefabrication areas?

    C. What off-site storage sites are available to handle items that cannot be stored on-site?

    D. What special facilities and equipment will be required for materials handling? Which are existing and which must be developed?

7. Utilities and temporary facilities

    A. Are existing structures available for construction use? If so, describe their potential use. If not, describe any development that is required. Obtain drawings or other dimensional data on these structures.

    B. May any of the permanent structures to be constructed be used by the contractor on a temporary basis during construction?

    C. Identify locations on the site or in nearby areas that are available for temporary structures. If off-site areas must be used, who provides them?

    D. Must the contractor provide temporary facility space for owner personnel or others?

    E. Will the owner take over any temporary structures when the project is complete or must the contractor remove them?

    F. Who provides office furnishings?

    G. Who provides potable water? What is its source and what development is required? Will there be separate water systems for concrete water, dust control, and fire protection? If so, what is their source, who provides them, and what development is required?

    H. What sanitary facilities (toilets, washrooms, sewer systems) exist? What must be developed and who provides them?

    I. How are heating and air conditioning handled and who provides them? What development is required?

    J. What is the availability of electrical power (location, voltage, phases, capacity)? What must be done to provide power and who provides it?

    K. Is steam required for construction use? If so, what is its source and who provides it?

    L. Is compressed air or other construction gas required for construction use? If so, what is its source and who provides it?

    M. Who provides trash and garbage pickup service?

8. Local materials, services, and subcontractors

    A. Provide names, addresses, and other data with respect to local specialty construction contractors whose services may be required in the performance of the contract.

    B. What is the source of ready-mix concrete and what is its price? If the contractor must supply concrete, what is the source of aggregates and cement?

    C. What trucking firms maintain offices in nearby communities?

    D. Obtain classified sections of telephone directories for nearby communities.

9. Local conditions

    A. If the site is within an operating facility, what special requirements exist with respect to gates, restricted areas, etc.? Does the potential exist for union problems between operating personnel and construction personnel?

    B. To what extent has the owner coordinated with local governments and agencies concerning this project?

    C. Do any political or cultural situations exist that can affect the project? Does the project enjoy public support?

    D. Are there any unresolved or incomplete actions (such as owner permits) that could delay the planned start of the project or affect construction progress?

    E. Is the project of such a nature and/or so located that adjacent communities are going to be subjected to high levels of construction noise, dust, traffic, or conditions that may precipitate complaints and legal actions or that otherwise adversely affect community relations?

10. Health, safety, environmental, and security
    A. What medical support will be available? Who provides it?
    B. Where are nearby hospitals, dispensaries, and clinics? What services do they have available?
    C. How is ambulance service provided? Is air evacuation service available?
    D. How will fire protection be provided?
    E. What is required to comply with environmental regulations relating to garbage, trash, dust, noise, erosion, hazardous waste, chemical spills, and burning? Who provides these services?
    F. Will project personnel be exposed to hazardous fumes, chemicals, or radiation for which special protective equipment and training must be provided?
    G. Is a local service available to provide chemical toilets? Are these the portable type or toilet trailers?
    H. What nearby commercial firms are licensed to handle and transport hazardous waste to licensed disposal areas?
11. Security
    A. What law enforcement agencies have jurisdiction over the project site, supporting communities, and transportation networks?
    B. If the project is open shop, will local law enforcement agencies provide protection in case of adverse union activity?
    C. Is security fencing to be provided? By whom? Who provides the guard force and any surveillance systems?
12. Transportation
    A. On a map showing the transportation network, identify:
       (1) The weight and dimensional limitations of any roads, bridges, and tunnels
       (2) Railroad lines and receiving docks
       (3) Available barge waterways and docks
       (4) Available airports
    B. For railroad service:
       (1) What railroad lines serve the area?
       (2) Will a railroad spur be required? If so, who provides it and what line does it tie into?
    C. Identify the commercial air carriers serving the area and the airports from which they operate. List the maximum size aircraft that can use each airport.
    D. For barge service, identify available dockage, any weight and dimensional restrictions, seasonal availability, or other restrictions. Also list other ports on the system.
    E. Describe public transportation available for personnel.

13. Labor
    A. If this is a union project, identify the names and location of locals whose crafts will be used. Get copies of all agreements and wage scales to be applied. Specifically identify any work rules involving limitations on production, overtime, travel pay and time, showup pay, etc. Note the expiration dates of all agreements and assess the potential for strikes upon expirations. Discuss any conditions relating to the area's union culture that may affect project performance and costs.
    B. If this is an open shop project, determine the prevailing wages through state employment agencies, local contractors, or other sources. Identify worker source areas and estimate the availability of each craft over the life of the project. In making these evaluations, consider the demands of other existing or planned projects in the area that may compete for the same personnel.
    C. Will any skills training of personnel be required? If so, which skills and who will provide the training?
    D. Evaluate the local economic situation with respect to its effect on availability, productivity, and wage scales for personnel.
    E. Describe any conditions (such as commuting distance or high crime area) that may limit personnel recruiting potential or contribute to high turnover.
    F. Obtain any data available that might show area labor productivity.
    G. Is there any potential in this area for government-subsidized job training programs for the work force?
14. Construction equipment
    A. Identify equipment that the owner or other contractors will provide. What is the reimbursement system for this support?
    B. Identify special types of equipment the contractor must provide.
    C. Will equipment servicing and maintenance be handled in on-site facilities or by contract with local firms?
    D. Will there be any impediments to moving equipment onto the job site (such as limited capacity roads and bridges, restrictive tunnels, or overhead wires)?
15. Communications
    A. How is telephone service to be provided?
    B. Will there be any on-site communications (such as loudspeaker or radio)? Who provides them?
    C. Is overnight express mail/parcel service available?
16. Permits, taxes, and fees
    A. Is the project tax exempt, either locally or federally?

    B. What state and local taxes will apply?

    C. Is a contractor license required to operate in the area? What are the requirements?

    D. What permits will be required of the contractor?

    E. What hookup fees will the contractor pay (phone, water, sewer, electrical, etc.)?

17. Professional and office staff

    A. Is adequate housing available in nearby communities for contractor professional staff?

    B. What are the availability and wage scale of local-hire office personnel?

18. Miscellaneous

    A. What support or services must the contractor provide to the owner, engineer, subcontractors, or others on site?

    B. Will participation in any special programs (such as substance abuse control) be required of the owner?

Project Leadership: Understanding and Consciously Choosing Your Style

INTRODUCTION

Trying to help project managers become better leaders has always been a difficult task. Project managers must rely on a variety of skills in order to successfully manage their projects. Not the least of these required skills is the ability to motivate, inspire, and lead the project team. True leadership on the part of the project manager has been shown time and again to be one of the most important single characteristics in successfully implementing projects.[1-3] As a result, the more one is able to understand about this complex concept, the better able one will be to manage one's own projects and train future project managers in the tasks and skills required for their jobs.

Leadership is a complex process that is crucial to successful project management. The principal difficulty with studying leadership is that we know at once so much and so little about the concept. Behavioral scientists have been studying

Reprinted from *Project Management Journal,* Vol. XXII, No. 1, pages 39–47, 1991 by Dennis P. Slevin and Jeffrey K. Pinto with permission of the Project Management Institute, 130 South State Road, Upper Darby, PA 19082, a worldwide organization of advancing the state-of-the-art in project management. Phone: (610) 734-3330, fax: (610) 734-3266.

the leadership problem over the past half century in an attempt to better understand the process and to come up with prescriptive recommendations concerning effective leadership behaviors. Years of careful research have generated a variety of findings that at best are sometimes confusing to the practicing manager and at worst are often internally inconsistent. For example, the following eleven propositions have all been suggested to guide leader behavior:

- Leaders will be most effective when they show a high level of concern for both the task and the employees (their subordinates).[4]
- Leaders should be task-oriented under conditions where they have either high or low control over the group. When they only have moderate control over the group, leaders should be people-oriented.[5]
- To be effective, leaders must base their decision-making style on how important employee acceptance of the decision will be. They should seek participation by subordinates in decision-making accordingly.[6]
- Leaders will be accepted and will motivate employees to the extent that their behavior helps employees progress toward valued goals and provides guidance or clarification not already present in the work situation.[7]
- A participative approach to leadership will lead to improved employee morale and increase their commitment to the organization.[8]
- Under stressful conditions (e.g., time pressure) an autocratic leadership style will lead to greater productivity.[9] A minimum level of friendliness or warmth will enhance this effect.[10]
- Even when a leader has all the needed information to make a decision and the problem is clearly structured, employees prefer a participative approach to decision-making.[11]
- Effective leaders tend to exhibit high levels of intelligence, initiative, personal needs for occupational achievement, and self-confidence.[12,13]
- Effective leadership behavior can depend on the organizational setting, so that similar behaviors may have different consequences in different settings.[14,15]
- Leadership behavior may not be important if subordinates are skilled, if tasks are structured or routine, if technology dictates the required actions, and if the administrative climate is supportive and fair.[16]
- Leadership behavior can be divided into task behavior (one-way communication) and relationship behavior (two-way communication). An effective leadership style using these behaviors can be selected according to the maturity level of subordinates relative to accomplishing the task.[17]

This list of principles of leadership and leader behavior presents a bewildering set of premises for project managers. In some cases, these points seem to actively disagree with each other. For example, some researchers say that a participatory leadership style is best for all situations, while other work suggests that a participatory style is more effective in some situations and with some types of subordinates than in others. Some of the conclusions argue that specific character traits of the leader determine effectiveness while other work states that it is, in fact, the dynamics of the interaction of the leader with subordinates that determines effectiveness. Finally, other authors have concluded that, under some conditions, leadership style is not important at all.

So where does this leave the project manager? These research results certainly reinforce the perception that the process of project leadership is confusing, complex, and often contradictory. What would help the practicing project manager is a day-to-day "working" model that clarifies some aspects of the concept of leadership, suggests alternative leadership styles that may be used, and provides some practical recommendations on conditions under which alternative leadership styles might be used. The purpose of this article is to describe a cognitive approach to leadership that will help the project manager consciously select the leadership style correct for alternative situations.

CRITICAL DIMENSIONS OF INFORMATION AND DECISION AUTHORITY

Suppose that you are a leader of a five-person project team. You are faced with a problem that is complex, yet a decision must be made. From the standpoint of leadership, before you make the decision, you must answer two "predecisional" questions:

1. **Where do you get the information input?** (Whom do you ask for relevant information?)
2. **Where should you place the decision authority for this problem?** (Who makes the decision?)

The first question asks which members of the group you head will furnish information about a particular decision. The second asks to what extent you maintain all the decision authority and make the decision yourself or to what extent you "share" your decision authority with members of your project team and have them make the decision with you in a more or less democratic fashion.

The first dimension is one of information; the second, one of decision authority. These two critical dimensions are essential for effective leadership and they have been plotted on the graph in Figure 1, The Bonoma/Slevin Leadership Model. As a leader, you may request large amounts of subordinate information

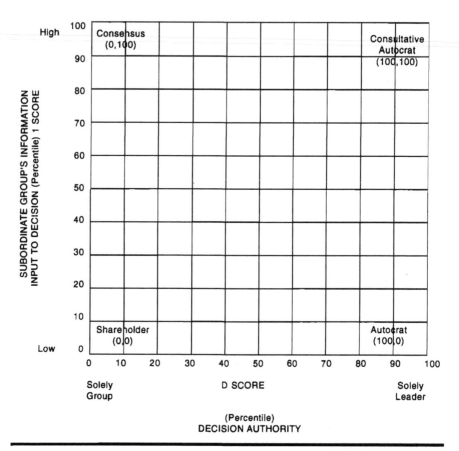

FIGURE 1 The Bonoma/Slevin Leadership Model.

input into a decision or very small amounts. This is the vertical axis — information input. Secondly, as a leader, your decision making style may range all the way from you making the decision entirely yourself to you sharing power with the group and having the final decision made entirely as a group decision. This is the horizontal axis — decision authority. (Note: The complete Jerrell/Slevin Leadership Model involves a series of questions that lead to percentile scores for plotting onto the above chart.) A copy of the Jerrell/Slevin Leadership Instrument can be obtained from the authors of this article.

By using percentile scores, any leadership style may be plotted in the two-dimensional space. For convenience, we will refer to the horizontal axis (decision authority) first and the vertical axis (information) second in discussing scores.

FOUR LEADERSHIP STYLES

Using the plotting system, we can describe almost every leadership style. (Refer to Figure 1.) The four extremes of leaders you have known (depicted in the four corners of the grid) are the following:

Autocrat (100,0). Such managers solicit little or no information from their project team and make the managerial decision solely by themselves.

Consultative autocrat (100,100). In this managerial style, intensive information input is elicited from the members, but such formal leaders keep all substantive decision-making authority to themselves.

Consensus manager (0,100). Purely consensual managers throw open the problem to the group for discussion (information input) and simultaneously allow or encourage the entire group to make the relevant decision.

Shareholder manager (0,0). This position is literally poor management. Little or no information input and exchange takes place within the group context, while the group itself is provided ultimate authority for the final decision.

The advantages of the leadership model, apart from its practical simplicity, become more apparent when we consider three traditional aspects of leadership and managerial decision style:

- Participative management
- Delegative
- Personal and organizational pressures affecting leadership

PARTICIPATIVE MANAGEMENT

The concept of "participation" in management is a complex one with different meanings for different individuals. When we discuss the notion of participation with practicing project managers in a consulting setting, the following response is typical:

> "Oh, I participated with my subordinates on that decision — I asked each of them what they thought before I made the decision."

To the practicing manager, participation is often an **informational** construct, i.e., permitting sufficient subordinate input to be made before the hierarchical decision is handed down. On the other hand, when we discuss the concept of participation with academics, the following response is typical:

> "Managers should use participation management more often — they should allow their subordinates to make the final decision with them in a consensual manner."

To the academic, the concept of participation is often an issue of **power**, i.e., moving to the left on the Bonoma/Slevin model such that decision authority is shared with the group.

In actuality, participation is a **two-dimensional** construct. It involves both the solicitation of information and the sharing of power or decision authority. Neither dimension, in and of itself, is a sufficient indicator of a participative leadership style. Participative management requires equal emphasis on the project manager requesting information and allowing a more democratic decision-making dynamic to be adopted.

DELEGATION

Good managers delegate effectively. In doing so they negotiate some sort of compromise between the extreme of "abdication" — letting subordinates decide everything — and "autocratic management" — doing everything themselves. We have found the leadership model useful for managers as they delegate work to their subordinates. After exposure to the model, managers are often likely to be more explicit about the informational and decision authority requirements of a delegated task. For example, the project manager may say to his subordinate:

> "Get the information that you need from my files and also check with Marketing (information). You make the decision yourself (decision authority), but notify me in writing as soon as you have made it so that I am kept informed."

Thus the subordinate understands both the information and decision authority aspects of the delegated tasks. Delegation often fails when the communication is unclear on either or both of these dimensions.

PRESSURES AFFECTING LEADERSHIP

As a project manager, you may act differently as a leader under different conditions, depending upon three kinds of pressure:

1. Problem attributes pressures
2. Leader personality pressures
3. Organization/group pressures

Think of the leadership model in terms of a map of the United States. Table 1 summarizes these pressures on leadership style in terms of geographical direc-

TABLE 1   Three Leadership Style Pressures

Type of pressure	Direction of pressure on leadership grid
Problem attributes pressures	
Leader lacks relevant information; problem is ambiguous.	North: more information needed.
Leader lacks enough time to make decision adequately.	South and east: consensus and information collection take time.
Decision is important or critical to leader.	North and east: personal control and information maximized.
Problem is structured or routine.	South and east: little time as possible spent on decision.
Decision implementation by subordinates is critical to success.	West and north: input and consensus required.
Leader personality pressures	
Leader has high need for power.	East: personal control maximized.
Leader is high in need for affiliation, is "people oriented."	North and west: contact with people maximized.
Leader is highly intelligent.	East: personal competence demonstrated.
Leader has high need for achievement.	East: personal contribution maximized.
Organizational and group pressures	
Conflict is likely to result from the decision.	North and west: participative aspects of decision-making maximized.
Good group–leader relations exist.	North and west: group contact maximized.
Centrality: formalization of organization is high.	South and east: organization style matched.

tion (for example, a movement "north" is a movement upward on the vertical axis of the leadership model, and so on).

Problem Attributes

Problem attribute pressures generate eastward forces in the leader. In other words, they often cause leaders to take the majority of or sole decision-making authority on themselves. This is especially true when problems are characterized as:

- Time bound (a quick decision is required)
- Important
- Personal
- Structured and routine

In such cases, it is very tempting to take "control" over the decisional process personally and "get the job done." However, northward pressures (involving the quest for additional information) can occur as well, given:

- Important decisions
- Decisions in which you as the leader lack the resources to make the decision yourself
- Problems in which subordinate implementation is critical to success

In these cases, information input and exchange will be maximized. Time pressures tend to push one toward the east, into an "autocratic mode." Implementation concerns push one to the northwest, in search of both more information and the active participation in and support of the decision by subordinates. It is a well-known behavioral finding that people cooperate more in implementing decisions they have helped make.

Leader Personality

Some managers tend to be inflexible in their leadership style because of who they are and how they like to manage. Our experience has led us to conclude that, in many cases, such managers:

- Have a high need for power
- Are task oriented, or
- Are highly intelligent

Managers of this type will make many decisions themselves that might otherwise be left to subordinates, and they also may make decisions without acquiring information from the group.

People-oriented leaders, on the other hand, will act to maximize information inputs from their subordinates and to share their decision authority as well. Both of these activities are "people processes."

Organizational/Group Pressures

If conflict is likely to result from any decision made, effective managers are most likely to want their subordinates as involved as possible in both the input (northward) and authority (westward) dimensions, so as to help manage the potential conflict. Should good leader/group relations exist, pressure northward (but NOT necessarily westward) will be felt. The leader will feel great pressure to fit into the "culture" of the organization. The R&D lab expects a consensual approach; most business groups expect a consultative autocrat approach; the factory floor may expect an autocratic approach. It is important to match your style to the norms, needs, and expectations of your subordinates.

Flexibility

We argue for a contingency model of leadership. No one style is best for all situations. Rather, the successful manager is flexible and moves around the model as appropriate to the given situation. If time pressures are overwhelming, autocratic decisions may be appropriate. For a majority of management decision making, the consultative autocrat approach may be the best alternative. And in dealing with the high-tech professionals, engineers, and other specialists, the project manager may choose a more consensual style. The key to success is to **match your leadership style to the situation.**

EXAMPLES OF THE MODEL IN USE

One of the author's Executive MBA students, with significant project management responsibility and experience, recently applied the Bonoma/Slevin model to a number of day-to-day leadership situations he faced. His comments and analysis are shown in Table 2. These examples clearly show flexibility (the willingness to move around the leadership model on a decision-to-decision basis) and self-insight.

TABLE 2   Case Examples for One Project Leader

1. PROJECT: Determine if the companies should take cash discounts where economically feasible and ensure these discounts are passed through to the customer.
Leadership Style: Autocrat (100,0)
Basis for Leadership Style: The question leads itself to a logical solution (which I am qualified and have the authority to make). Implementation is primarily focused on adequate contacts versus a group consensus/commitment.

2. PROJECT: Initiate a study to consider involving the salesmen in overall company profitability.
   a. Should the formula for salesmen's compensation take into consideration the level of gross margin achieved on each sale?
   b. What expenses are controlled by salesmen? How should these expenses be monitored?
   c. Should salesmen be involved in the collection of accounts receivable?
Leadership Style: Consultative Autocrat (80,80)
Basis for Leadership Style: Similar to project 1, this question represents a financial control decision to the organization. Implementation of the decision can be performed by focusing on adequate controls and does not require a group consensus. Unlike project 1, I do not feel qualified to autocratically make this decision. It requires a detailed knowledge of current procedures, industry "normal," and individual corporate cultures. Accordingly, considerable consultation and data collection will have to be performed prior to making and implementing a decision.

3. PROJECT: Coordinate efforts when a new product is introduced and manufacturer support is not adequate, to assure salesmen and service training is appropriate.
Leadership Style: Consensus Manager (10,90)
Basis for Leadership Style: The subject is one which requires information from the group as well as a group commitment for successful implementation. The group process will likely generate various alternatives which will be incorporated into the final decision. Equally important as the leadership style (consensus manager) is the composition of the group. Accordingly, members of management, sales, and several technicians will be asked to participate in this process.

4. PROJECT: Investigate combined purchases within zones and throughout the country.
Leadership Style: Consensus Manager (30,80)
Basis for Leadership Style: This decision process requires a consensus approach primarily due to the need for group information, support, and implementation. Implementation will require a joint effort of the group with respect to initial purchasing procedures as well as obtaining the manufacturer's approval for combining purchases across regions. Said approval has historically been denied. The probability of such an agreement from the manufacturer is enhanced if approached by a group which has the ability to capitalize on its size and existing relationships.

5. PROJECT: Develop and circulate monthly listings of each company's slow moving inventory.

TABLE 2   Case Examples for One Project Leader (continued)

Leadership Style: Mix between Consultative Autocrat and Consensus Manager (50,70)
Basis for Leadership Style: The decision to circulate the lists could be purely autocratic; however, proper implementation will require group acceptance. Therefore, it is important to involve the group. I believe the group will come to the desired outcome. If, however, it becomes apparent that the group does not adequately assess the situation, a shift to a consultative autocrat approach will be implemented due to the underlying importance of this project. (Note: Severity of the problem is exemplified by the existence of inventory days, at certain locations, in excess of 300 days in an industry where inventory days average 65–75 days).

6. PROJECT: Determine the need for EDP systems coordination, enhancement, and consolidation.
Leadership Style: Mix between Autocrat and Consensus Manager; 6a (70,30) and 6b (20,70)
Basis for Leadership Style: This style initially appears contradictory. However, I view this objective as two separate decisions:

   a. How much can be spent (i.e., based on appropriate cost/benefits and risk analysis)?
   b. What systems should be selected and implemented based on financial guidelines?

Accordingly, I believe the decision in step one (part a) can be made in an autocratic style based on the investor's returns and "value added" requirements while the actual system and method of implementation can be decided upon as a group consensus. The key to the above project is to explain to the project team the critical success factors and parameters existing in Decision 1 and their inherent limitations for Decision 2.

7. PROJECT: Develop standardized financial reporting.
Leadership Style: Autocrat (70,30)
Basis for Leadership Style: The desired result is a highly structured financial report among each individual organization which can be utilized for comparison and consolidation purposes. Accordingly, there is virtually no room for flexibility. A limited amount of consulting may be performed to enhance the investor's understanding of the industry. It is highly unlikely that this will have an impact on the report structure. (It is important to note that additional information applicable to the industry may be supplementally obtained, but the primary report will be consolidated and compared with other investments. Hence, flexibility is severely limited.)

8. PROJECT: Coordinate efforts to reduce parts inventory.
Leadership Style: Mix between Consultative Autocrat and Consensus Manager (50,70)
Basis for Leadership Style: Similar to project 5. A consultative autocrat approach may be necessary to stress the point that something must be done. Yet, a consensus manager approach is necessary to obtain information, utilize the appropriate expertise, and ensure implementation.

9. PROJECT: Develop a joint program to dispose of used equipment.
Leadership Style: Mix between Consultative Autocrat and Consensus Manager (50,70)

TABLE 2   Case Examples for One Project Leader (continued)

Basis for Leadership Style: This project is designed to dovetail on project 5. I believe the same leadership style is appropriate.

10. PROJECT: Determine the feasibility and cost benefits of joint advertising and telemarketing.
Leadership Style: Consensus Manager (30,90)
Basis for Leadership Style: Implementation of this project requires group commitment. If the group decides that joint advertising and telemarketing are beneficial, then considerable planning and coordination will be necessary for project success.

Case Examples for One Project Leader

The following is the analysis of one project manager in a responsible position using the model to analyze ten decisions as described in his own words.[18]

> I have elected to analyze the decision process/leadership style I could utilize in ten key projects associated with the assemblage discussed in the introduction.
>
> First, I completed the Jerrell/Slevin Management Instrument. This instrument identified my decision process to be "balanced" in that it is not heavily skewed to any one of the four management styles. The measurement tended to lean toward consultative autocrat.
>
> Second, I asked my superiors as well as my subordinates and peers to categorize my decision/leadership style as autocrat, consultative autocrat, consensus manager, or shareholder manager based on their definitions as previously discussed. Almost unanimously, these individuals categorized my style as Consultative Autocrat.
>
> Being consciously aware of the above and the fact that an effective leader must have a leadership style which is flexible to individual situations, I have listed the projects, identified what I feel to be the appropriate leadership style, and estimated the coordinates of this leadership style on the Bonoma/Slevin Leadership Model. They are as in Table 2, Case Examples for One Project Leader.

FINDINGS FROM PERSONAL EXPERIENCE

Based on the presentation and discussion of this leadership model with thousands of practicing managers, we would like to share with you some of our conclusions concerning project management and leadership style.

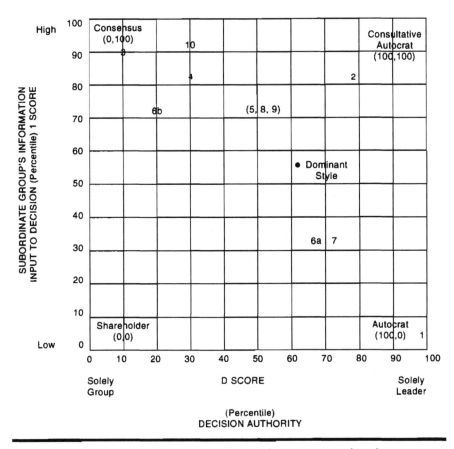

FIGURE 2   Bonoma/Slevin Leadership Model with ten projects plotted.

You are more autocratic than you think.   In the eyes of your subordinates, you are probably closer to the autocrat on the graph than you are in your own eyes. Why? Because you are the boss and they are the subordinates. No matter how easygoing, friendly, participative, and supportive you are, you are still the boss. There is almost always a difference in leadership style as perceived by supervisor and subordinates.

But it's O.K.   Often, when you ask subordinates where they would like their boss to move, they respond, "Things are O.K. as they are." Even though there are perceptual discrepancies concerning leadership style, there may not necessarily be a felt need or pressure for change. The status quo may be O.K.

It is easy to move north and south. It is easy to move vertically on the graph. Why? Because management is a job of communication. It is easy to collect more information or less information before you make the decision. The information dimension is the least resistant to change.

It is hard to move west. Most managers, in our experience find it quite threatening to move westward too quickly. This hesitation is because a westward movement upsets the basic power realities in the organization. If your head is in the noose, if things do not work out, then it is hard to turn the decisions over to your subordinates.

If subordinates' expectations are not met, morale can suffer. What would happen if your subordinates expected you to use a (50,90) process and instead you made the decision using a (90,10) style? That's right, dissatisfaction and morale problems. As mentioned before, decision process can be as important as decision outcome, especially from the standpoint of motivating subordinates.

Be flexible. A successful manager is autocratic when he/she needs to be, consultative when necessary, and consensual when the situation calls for it. Move around the leadership space to fit the needs of the situation. Unsuccessful managers are inflexible and try the same style in all situations. Most managers feel that their score on the Jerrell/Slevin Management Instrument is a function of the particular job they have been in over the last few months. Be flexible and match your leadership style to the job.

OVERVIEW: IMPLICATIONS FOR YOUR LEADERSHIP STYLE

The leadership framework presented in our model forces the manager to ask two key questions concerning decision making:

- Whom do I ask?
- Who makes the decision?

Obtaining accurate information input from one's subordinate group is crucial to effective project management. Similarly, decision-making authority must be located in the right place vis-a-vis the leader's group.

The model presented here might be more broadly applied to projects in general. Successful projects are implemented through sufficient access to and judicious use of both information and power. At the start of any project one might be well advised to review the status of project information and power. Consider the following questions:

- Where will the project information come from?
- Does the project team have adequate access to information
  - From top management
  - From peers
  - From technical experts
  - From government regulators
  - From the formal MIS
  - From other team members
- Who has the power to make decisions and get those decisions implemented?

Once the team has developed technical/analytical solutions, who has the organizational authority to get action? Do the project leaders have the decision authority they need to command resources, people, material, and behavior change? If some project leaders do not have formal power, as can sometimes be the case, do they have access to top management or other organizational sources of power? In summary, successful project management is greatly affected by the information and power in the system. Successful project managers have always had an intuitive sense for these issues. This model is an attempt to provide the project manager with a framework that can be used in a conscious and analytical way to enhance project success.

ACKNOWLEDGMENT

The authors are indebted to Thomas V. Bonoma for the development of the Bonoma/Slevin Leadership Model and to S. Lee Jerrell for the development of the Jerrell/Slevin Management Instrument.

NOTES

Portions of this article were adapted from the chapter on Leadership in *The Whole Manager: How to Increase Your Professional and Personal Effectiveness,* by Dennis P. Slevin, AMACOM: New York, 1989, © 1989 by Dennis P. Slevin. Used with permission, and "Leadership, Motivation and the Project Manager," by Dennis P. Slevin and Jeffrey K. Pinto in *Project Management Handbook,* D.I. Cleland and W.R. King (Eds.), Van Nostrand Reinhold Co.: New York, 1988.

REFERENCES

1. Slevin, D.P. and Pinto, J.K. 1988. Leadership, Motivation, and the Project Manager. In *The Project Management Handbook,* 2nd Ed., ed. D.I. Cleland and W.R. King, 739–770. New York: Van Nostrand Reinhold Co.

2. Baker, B.N., Murphy, D.C. and Fisher, D. 1983. Factors Affecting Project Success. In *The Project Management Handbook,* ed. D.I. Cleland and W.R. King, 69–85. New York: Van Nostrand Reinhold Co.

3. Posner, B.Z. 1987. What It Takes to Be a Good Project Manager. *Project Management Journal,* vol. XVIII, no. 1, 51–54.

4. Blake, R.R. and Mouton, J. 1964. *The Managerial Grid.* Houston, TX: Gulf Publishing.

5. Fiedler, F.E. 1978. Contingency Models and the Leadership Process. In *Advances in Experimental Social Psychology,* ed. L. Berkowitz, vol. 11. New York: Academic Press.

6. Vroom, V.H. and Yetton, P.W. 1973. *Leadership and Decision Making.* Pittsburgh, PA: University of Pittsburgh Press.

7. House, R.J. 1971. A Path–Goal Theory of Leadership Effectiveness. *Administrative Science Quarterly,* vol. 16, 321–333.

8. Fleishman, E.A. 1973. Twenty Years of Consideration and Structure. In *Current Developments in the Study of Leadership,* ed. E.A. Fleishman and J.G. Hunt. Carbondale, IL: Southern Illinois University Press.

9. Fodor, E.M. 1976. Group Stress, Authoritarian Style of Control, and Use of Power. *Journal of Applied Psychology,* vol. 61, 313–318.

10. Tjosvold, D. 1984. Effects of Leader Warmth and Directiveness on Subordinate Performance on a Subsequent Task. *Journal of Applied Psychology,* vol. 69, 422–427.

11. Heilman, M.E., Hornstein, H.A., Cage, J.H. and Herschlag, J.K. 1984. Reactions to Prescribed Leader Behavior as a Function of Role Perspective: The Case of the Vroom–Yetton Model. *Journal of Applied Psychology,* vol. 69, 50–60.

12. Stodgdill, R. 1984. *Handbook of Leadership,* New York: Free Press.

13. Ghiselli, E. 1971. *Exploration of Managerial Talent,* Santa Monica, CA: Goodyear.

14. Dansereau, F., Graen, G., and Haga, B. 1975. A Vertical Dyad Linkage Approach to Leadership Within Formal Organizations: A Longitudinal Investigation of Role Making Process. *Organizational Behavior and Human Performance,* vol. 13, 45–78.

15. Salancik, G.R., Calder, B.J., Rowland, K.M., Leblibici, H., and Conway, M. 1975. Leadership as an Outcome of Social Structure and Process: A Multi-Dimensional Analysis, In *Leadership Frontiers,* ed. J.G. Hunt and L.. Larson, Kent, OH: Kent State University Press.

16. Kerr, S. and Jermier, J.M. 1978. Substitutes for Leadership: Their Meaning and Measurement. *Organizational Behavior and Human Performance,* vol. 22, 375–403.

17. Hersey, P. and Blanchard, K.H. 1977. *Management of Organizational Behavior: Utilizing Human Resources,* 3rd Ed. Englewood Cliffs, NJ: Prentice-Hall.

18. Sprecher, W.E. April 1987. Private Communication.

APPENDIX E

Work Breakdown Structure Templates

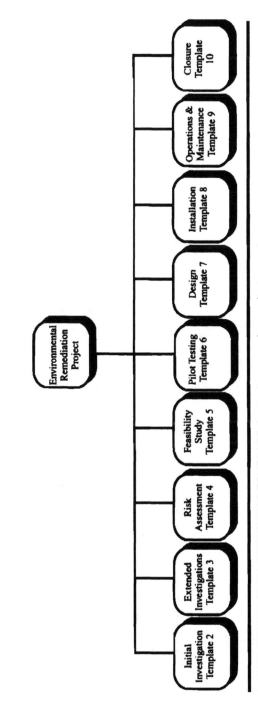

WORK BREAKDOWN STRUCTURE TEMPLATE 1 Environmental remediation project stages.

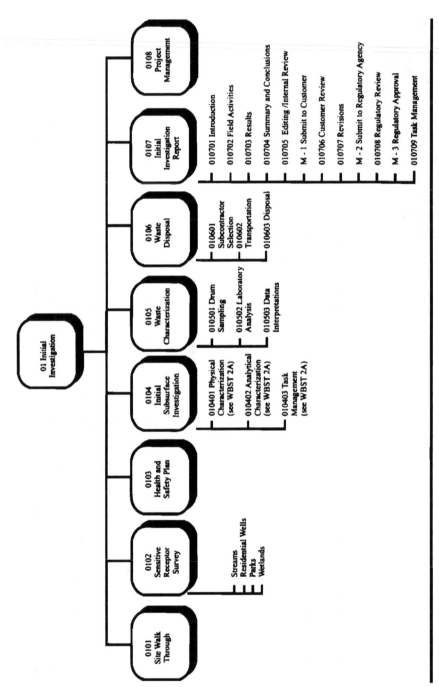

WORK BREAKDOWN STRUCTURE TEMPLATE 2 Initial investigation.

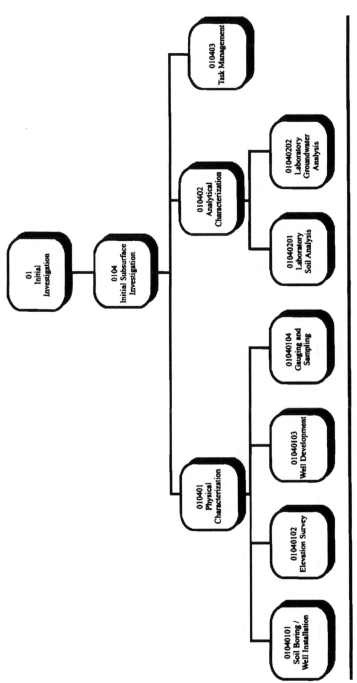

WORK BREAKDOWN STRUCTURE TEMPLATE 2A Initial investigation.

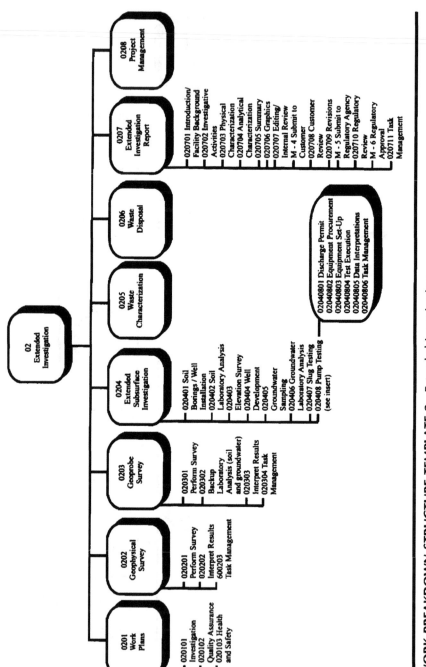

WORK BREAKDOWN STRUCTURE TEMPLATE 3 Extended investigation.

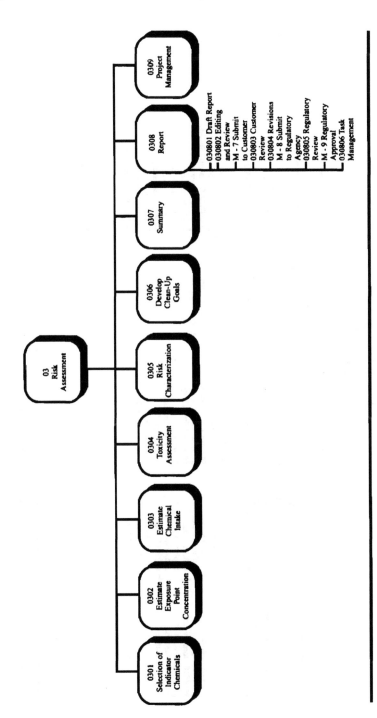

WORK BREAKDOWN STRUCTURE TEMPLATE 4 Risk assessment.

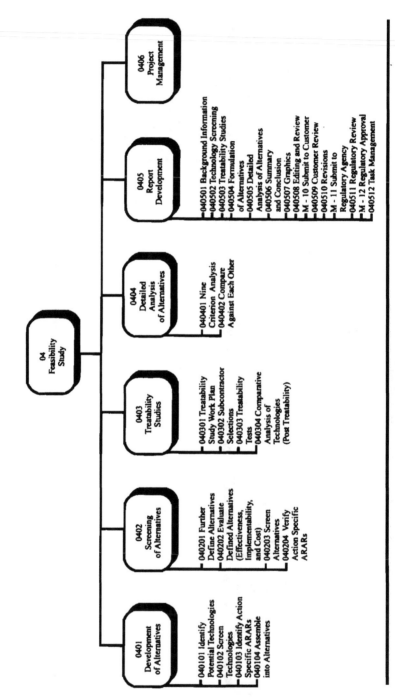

WORK BREAKDOWN STRUCTURE TEMPLATE 5 Feasibility study.

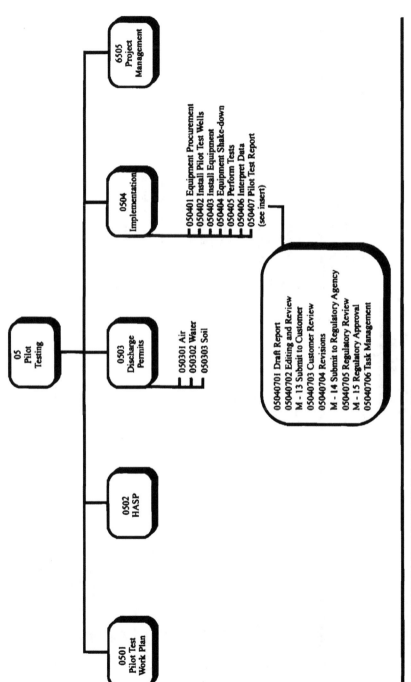

05 Pilot Testing

0501 Pilot Test Work Plan

0502 HASP

0503 Discharge Permits
- 050301 Air
- 050302 Water
- 050303 Soil

0504 Implementation
- 050401 Equipment Procurement
- 050402 Install Pilot Test Wells
- 050403 Install Equipment
- 050404 Equipment Shake-down
- 050405 Perform Tests
- 050406 Interpret Data
- 050407 Pilot Test Report (see insert)

6505 Project Management

- 05040701 Draft Report
- 05040702 Editing and Review
- M - 13 Submit to Customer
- 05040703 Customer Review
- 05040704 Revisions
- M - 14 Submit to Regulatory Agency
- 05040705 Regulatory Review
- M - 15 Regulatory Approval
- 05040706 Task Management

WORK BREAKDOWN STRUCTURE TEMPLATE 6   Pilot testing.

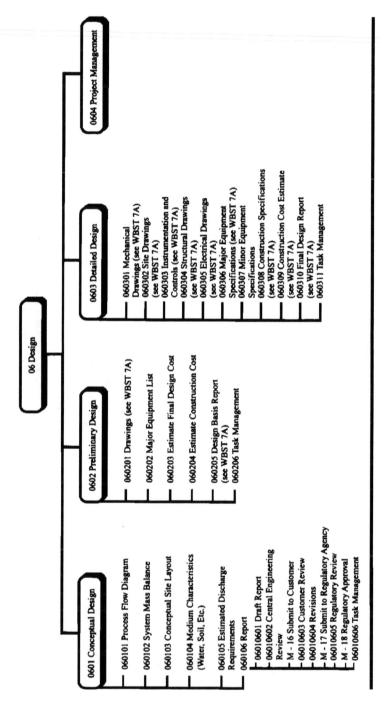

06 Design

0601 Conceptual Design

- 060101 Process Flow Diagram
- 060102 System Mass Balance
- 060103 Conceptual Site Layout
- 060104 Medium Characteristics (Water, Soil, Etc.)
- 060105 Estimated Discharge Requirements
- 060106 Report
  - 06010601 Draft Report
  - 06010602 Central Engineering Review
  - M - 16 Submit to Customer
  - 06010603 Customer Review
  - 06010604 Revisions
  - M - 17 Submit to Regulatory Agency
  - 06010605 Regulatory Review
  - M - 18 Regulatory Approval
  - 06010606 Task Management

0602 Preliminary Design

- 060201 Drawings (see WBST 7A)
- 060202 Major Equipment List
- 060203 Estimate Final Design Cost
- 060204 Estimate Construction Cost
- 060205 Design Basis Report (see WBST 7A)
- 060206 Task Management

0603 Detailed Design

- 060301 Mechanical Drawings (see WBST 7A)
- 060302 Site Drawings (see WBST 7A)
- 060303 Instrumentation and Controls (see WBST 7A)
- 060304 Structural Drawings (see WBST 7A)
- 060305 Electrical Drawings (see WBST 7A)
- 060306 Major Equipment Specifications (see WBST 7A)
- 060307 Minor Equipment Specifications
- 060308 Construction Specifications (see WBST 7A)
- 060309 Construction Cost Estimate (see WBST 7A)
- 060310 Final Design Report (see WBST 7A)
- 060311 Task Management

0604 Project Management

WORK BREAKDOWN STRUCTURE TEMPLATE 7 Design.

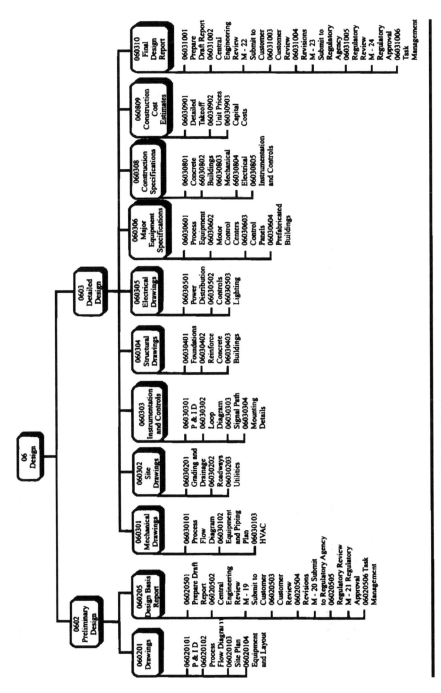

WORK BREAKDOWN STRUCTURE TEMPLATE 7A Design.

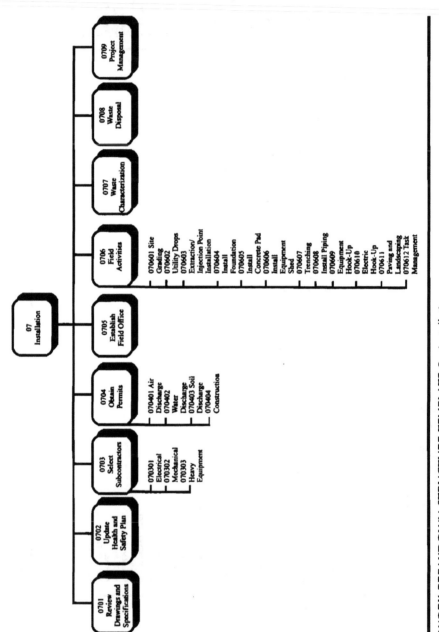

WORK BREAKDOWN STRUCTURE TEMPLATE 8 Installation.

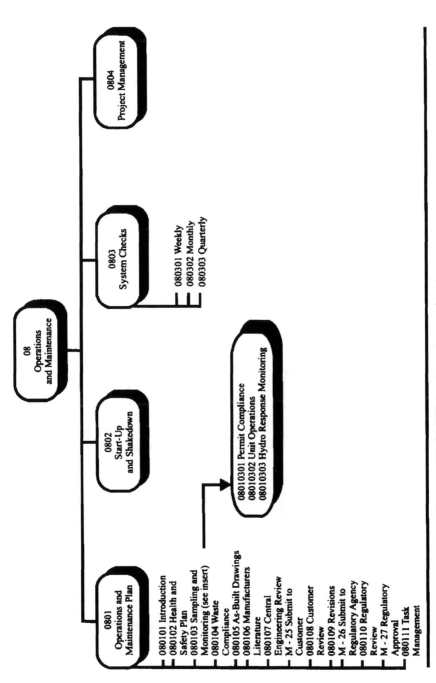

08
Operations and Maintenance

0801 Operations and Maintenance Plan

- 080101 Introduction
- 080102 Health and Safety Plan
- 080103 Sampling and Monitoring (see insert)
- 080104 Waste Compliance
- 080105 As-Built Drawings
- 080106 Manufacturers Literature
- 080107 Central Engineering Review
- M - 25 Submit to Customer
- 080108 Customer Review
- 080109 Revisions
- M - 26 Submit to Regulatory Agency
- 080110 Regulatory Review
- M - 27 Regulatory Approval
- 080111 Task Management

- 08010301 Permit Compliance
- 08010302 Unit Operations
- 08010303 Hydro Response Monitoring

0802 Start-Up and Shakedown

0803 System Checks

- 080301 Weekly
- 080302 Monthly
- 080303 Quarterly

0804 Project Management

WORK BREAKDOWN STRUCTURE TEMPLATE 9   Operations and maintenance.

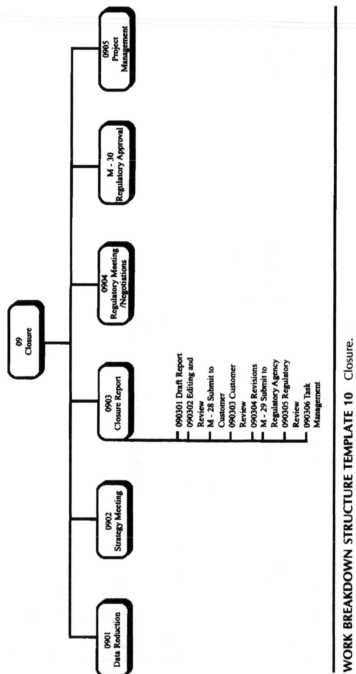

WORK BREAKDOWN STRUCTURE TEMPLATE 10 Closure.

Answers to
Chapter Exercises

Bid - No Bid Evaluation Form Page 1 of 2

A. Milestones

Date RFP Issued: 2/5/98	Date RFP Received: 2/5/98	Bid Decision Date: 3/5/98	Date Proposal Due: 2/20/98

B. Proposal Information

Lead Person: Frank Jones   Proposal No.: 01743   Customer: MegaCorp   Customer Contact: Tom Turbine
Telephone No.:   Location: Project Name: Former compressor plant
Project Description: Investigation/feasibility study/pilot/design
Competitors: 2
Special Considerations: pre-sold HVDPE   Unique Company Qualifications:
Last Visited By: sales staff   Visit Date: 11/5/97

C. Bid - No Bid Decision

	Matrix Weighted Decision Criteria Negative 0-3	Neutral 4-6	Positive 7-10	Your Company	Competitor A	Competitor B	Total Estimated Score / Maximum Score Available (# of factors X 10)
1. Previous experience with customer?	Poor	Average	(Good)	5	4	4	
2. Previous experience with similar projects?	Poor	Average	(Good)	7	5	4	
3. Opportunity was pre-sold by marketing?	No	No	(Yes)	10	7	8	
4. Alignment with your company's business plan?	Poor	Fair	(Good)	10	10	10	
5. Pace competition/number of competitors?	Strong	Medium	(Weak)	6	4	4	
6. Are we bidding against an incumbent?	Yes	Maybe	(No)	10	7	6	
7. Conflict of interest?	Yes	Possibly	(No)	10	10	10	
8. Availability of required operational resources?	Poor	Average	(Good)	8	6	6	
9. Availability of technical expertise?	Poor	Average	(Good)	10	4	4	
10. Customer's financial condition, stability, and reputation?	Weak	Medium	(Strong)	10	10	10	
11. Anticipated follow - on work?	No	Possibly	(Yes)	5	5	5	
12. Company office location?	Unfavorable	Neutral	(Favorable)	10	5	5	
13. Rate schedule/profitability?	Unfavorable	Neutral	(Favorable)	10	6	8	
14. Potential for new technology or new markets?	Poor	Average	(Good)	10	10	10	
15. Does customer have funds approved?	No	Don't Know	(Yes)	10	10	10	
16. Does the project require regulatory approval?	Yes	Don't Know	(No)	10	10	10	
17. If yes, do they have approval?	No	Don't Know	(Yes)	10	10	10	
				155	116	119	

Comments: *This is the type of project that we want.*

Prepared by: Joe Smith   Date: 2/6/98   Reviewed by: Sue Jones   Date 2/6/98   Preliminary Decision:
Prepared by: Reviewed by: Date: (Bid)   No Bid

D. Intermediate (Contract) Bid - No Bid Decision

Key Contract Evaluation Criteria   Circle Appropriate Rating

18. Terms and Conditions	Negative	(Neutral)	Positive
19. Insurance and Bonding Requirements	Unfavorable	(Average)	Favorable
20. Significant Contract Negotiation Required	Not Acceptable	(Some)	Acceptable
21. Level of Potential Risk (including 3rd party)	Yes, Not Allowed	High / (Average)	No or Not Required / Low

Comments:

Contract Type (circle one):   (Firm Fixed Price)   Time & Materials   Cost Plus Fixed Fee
Intermediate Decision (circle one):   (Bid)   No Bid
Contract Review: Frank Miller   Date: 2/7/98

RFP Summary Form	Page 1 of 7
Background Information	
Operational History	

Known	*Unknown*
⇒ Air compressor manufacturing plant in operation from early 1950s until 1987	⇒ Inventory of chemicals used in manufacturing process
⇒ Manufacturing building and 20 acres	⇒ Full description of manufacturing process
⇒ Best Auto Parts uses the former manufacturing building for warehousing only	⇒ Location of former disposal pits
⇒ 1985 leak occurred in 12,000 gallon UST	⇒ Have leaks occurred inside building
⇒ 1986 UST removed, excavation filled with crushed limestone	
⇒ All process machinery, equipment and supplies owned by Megacorp removed prior to sell	
⇒ Electrical sub-station, northeast side of lot has transformer filled with PCBs	
⇒ Property north and east of the parking area still owned by Megacorp	
⇒ Megcorp's property is referred to as the former disposal area	
⇒ Paints, miscellaneous fluids, parts, old tools, etc. are known to be deposited in four disposal pits	

RFP Summary Form	Page 2 of 7
Background Information	
Current Site Conditions	

Known	*Unknown*
⇒ Eighteen monitoring wells installed during previous environmental investigations.	⇒ Condition of monitoring wells
⇒ Three monitoring wells installed in former disposal area.	
⇒ Seven recovery wells installed in former UST area, one inch diameter, 9 feet in depth; recovery wells are not functional.	

RFP Summary Form	Page 3 of 7
Background Information	
Previous Environmental Activities (if any)	

Known	*Unknown*
⇒ Envirofixx, Inc. (EFX) removed leaking 12,000 gallon xylene tank; depth of excavation - 13.5 feet (1986)	⇒ Well schematics
	⇒ Locations of historical disposal trenches used by plant
⇒ EFX installed a total of six monitoring wells in the vicinity of the excavation. Four wells are considered shallow (screened between 5 to 10 feet) and two wells are considered deep (12 to 18 feet)	⇒ Actual locations of Terra Firma excavations
	⇒ Did tank excavations penetrate both sand layers
⇒ Terra Firma Corporation (TFC) investigated former disposal area by installing ten trenches and collecting 12 soil/waste samples (PCB analysis only). TFC logged the presence of soil staining, drum deris, paint sludge. clinker, etc. (Early 1986)	
⇒ In July of 1986 Air, Land & Sea Associates (ALS) installed six shallow monitoring wells (MW-1 through MW-6; 5 to 10 feet). ALS also installed a deep monitoring well, MW-7, 100 feet south of the former UST (12 to 18 feet). A total of four soil samples were collected and analyzed for ethylbenzene and xylene (MW-5 and MW-6).	
⇒ Eco-Pure, Inc. (1990) installed six monitoring wells (MW-8 to MW-13) and excavated 13 test pits in the former disposal area.	
⇒ Eco-Pure collected groundwater samples from all wells and analyzed them for TAL volatiles and PCBs.	
⇒ Monitoring wells MW-11, MW-12, and MW-13 are located in the former disposal area. Soil samples were collected from these wells and analyzed for pesticides, PCBs, TCLP metals, TAL volatiles, and TAL semi-volatiles.	

RFP Summary Form	Page 4 of 7

Physical Characterization

Regional Geology

Known	*Unknown*
⇒ Glacial till, clay, sand, and gravel from surface to approximately 140 feet. ⇒ Regional groundwater flow is to the southeast toward the Lower Takoma River.	

Site Geology

Known	*Unknown*
⇒ Geology consists of sand and gravel lenses which are discontinuous laterally and separated vertically by clay ⇒ Two water bearing lenses have been reported in the vicinity of the former UST. The shallower zone exists within a depth of 8-10 feet. The second zone exists within a depth of 12 to 14 feet. ⇒ A saturated sand lens has been reported in the former disposal area at a depth of 12 to 14 feet. ⇒ Groundwater flow appears radial from backfilled excavation.	⇒ If the sand lens at 12-14 feet in the UST area is in communication with the sand lens located at 12-14 feet in the former disposal area

RFP Summary Form	Page 5 of 7
Analytical Characterization	
Soil Analytical Results	

Known	*Unknown*
⇒ Highest PCB concentration TR-12 depth of six feet - 68 ppm ⇒ Highest xylene concentration in soil 3050 ppm, MW-5; 5 feet to 6.5 feet.	⇒ Background concentrations of total metals in soil ⇒ Location of trench and test pit samples

Groundwater Analytical Results	

Known	*Unknown*
⇒ Groundwater contains primarily xylene, however, concentrations of benzene and PCBs have also been found. ⇒ Extent of groundwater impact has not been fully determined (see MW-4) ⇒ Historical analysis indicates that impact limited to former UST area and to the shallower of two saturated zones. ⇒ Wells last sampled in September 1990	⇒ Background concentrations of metals in water ⇒ Has groundwater impact moved off site?

Sediment Analytical Results	

Known	*Unknown*
⇒ No sediment samples collected	⇒ Have sediments been impacted by the site?

RFP Summary Form	Page 6 of 7
Regulatory Issues	

Known	Unknown
⇒ State recently developed voluntary clean-up program; applies only to sites without prior regulatory action	⇒ Clean-up concentrations to be achieved in soil and groundwater
⇒ MegaCorp would like to qualify for voluntary clean-up program	
⇒ EPA in process of trying to score site for NPL listing	
⇒ If listed, EPA would like to address site under SACUM process	
⇒ If site included on NPL list it cannot be addressed by voluntary clean-up program	
⇒ Xylene is a "U" listed hazardous waste (U239)	
⇒ TSCA regulations will be a factor	
⇒ MegaCorp would like to begin project without regulatory approval	
⇒ MegaCorp wants the bidding contractor to tell them if a quality assurance is needed	
⇒ Paul Stanley would like the site to become a Superfund site	
⇒ EPA may issue a consent order before the years end	
⇒ Historically, state concerned over PCBs, not xylene	
⇒ State air permits allow 20 lb. VOCs/day	
⇒ Can access sanitary sewer on-site, three months to obtain discharge permit	

RFP Summary Form	Page 7 of 7
Regulatory Issues	

Known	*Unknown*
⇒ Local office 3.5 hours from site	⇒ Did electrical transformer leak in the past?
⇒ Local office only has experience with retail petroleum projects	
⇒ **RFP Objectives** • Development and execution of a work plan for final investigation • Work plan to include pilot testing and design of appropriate remedial technology • A cost estimate for technology implementation not required at this time • Must provide cost for health and safety plan	
⇒ Only time and materials not to exceed contracts are permitted	
⇒ A standard contract has not been negotiated with MegaCorp	
⇒ MegaCorp looking for a logical approach and strong project management skills	
⇒ Tom Turbine would like to see HVDPE pilot tested	
⇒ Maureen Jones wants to test gigometer	
⇒ Best Auto Parts concerned about disruption to operations	
⇒ Group of laid off workers are preparing to sue MegaCorp	
⇒ Competition; Eco-Pure & SoilRem	
⇒ Eco-Pure is known for investigation and not remediation	
⇒ Principals of SoilRem are former associates of Maureen Jones	
⇒ SoilRem not skilled in investigation	

Stakeholder Management Summary Form

Stakeholder Name	Specific Stake	Strengths	Weaknesses
Maureen Jones	Environmental Manager-Must have 15 remediation systems installed this year, must demonstrate competence; cost control	Position, respected and experience	Reports to legal department may rush project
Paul Stanley	Maintain control; demonstrate competence; protect company's liability; keep options open	Position, legal knowledge	Overly cautious, lacks technical knowledge
Tom Turbine	Project manager-needs to get project completed on time and under budget, must be assured technology will work, may be an ally of EnviroServ	Position, technical ability, contacts	Extremely busy, difficult time keeping up with details
Bill Preston	Project Engineer-needs to impress Maureen, Paul, and Tom, primary point of communication between EnviroServ and MegaCorp	Technical ability, enthusiasm	Inexperience, caught between issues
MegaCorp Senior Management	Bottom line cost and liability, schedule pressure, reduce environmental liabilities	Position, control funds	Removed from details, late involvement
EnviroServ	Win project, demonstrate HVDPE (High Vacuum Dual Phase Extraction), additional experience, earn a profit, expand industrial market	Pre-sold, technical experts, pool of resources	Location of nearest office, experience of local personnel
Eco-Pure	Win project, earn profit, win back trust of MegaCorp	Investigation	Remediation technology
SoilRem	Demonstrate ultra Megavac, win project, expand market	Remediation	Investigation
State Regulatory Agency	Concern over PCBs, demonstrate voluntary clean-up program, must permit discharge to sanitary sewer	Mandate	May have to defer to EPA
EPA	Protection of environment, mandate scoring site for NPL list	Position	Slow operations
Best Auto Parts	Concern over site disruption	Can prevent access	Must ultimately allow clean-up
Laid Off Workers	Payment from MegaCorp, develop case	Organization	Lack of funds, technical basis

Page 1 of 2

Stakeholder Management Summary Form

Stakeholder Name	Expected Strategy or Behavior	Stakeholder Management Response Strategies
Maureen Jones	Anxious to get started; demonstrate competence, must show controlling cost, may want to hire SoilRem, difficulties with Paul Stanley	Strategy for rapid installation, play to cost containment, give a strategy to manage Paul Stanley
Paul Stanley	Will try to avoid commitments	Give strategy to slow up project and avoid commitments, make him feel in control
Tom Turbine	Will try to be certain that technology will work, interested in PM controls, may be an ally	Provide as much technical data as possible, demonstrate PROMYSIS, offer training course
Bill Preston	Will want to be informed, caught in the middle	Develop strategy to satisfy Paul and Maureen (present strategy to Bill first)
MegaCorp Senior Management	Will stay uninvolved until late in the project	Invite to project kickoff meeting
EnviroServ		
Eco-Pure	May try to underbid	Discuss life cycle cost in EnviroServ proposal
SoilRem	Will heavily market ultra Megavac	Discuss comprehensive approach in proposal
State Regulatory Agency	Will want sampling for PCBs, may organize own investigation	Inform of plans for voluntary clean-up, include PCBs in sampling program
EPA	May organize own investigation, may issue a consent order in several months	Apply for voluntary clean-up program ASAP
Best Auto Parts	Complain about site disruption	Stay in constant communication, invite to kick-off meeting
Laid Off Workers	Will file law suit, attempt to demonstrate harm	Market risk assessment and sampling, in support of risk assessment

Page 2 of 2

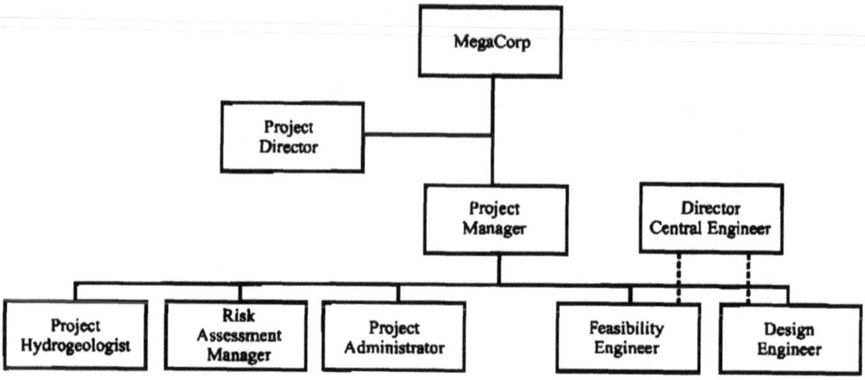

Project organization chart

Operations and Team Resources

Activities /Function	Project Director	Project Manager	Project Hydrogeologist	Director Central Engineer	Feasibility Engineer	Design Engineer	Risk Assessment Manager	Project Administrator	Operations Manager	Contract Administrator
Project Objectives	■	●	△	▲	△	△	△			
Preliminary Bid/No Bid	△	▲		▲						
Intermediate Bid/No Bid	△	▲							✖	●
Final Bid/No Bid	■	●	△	△	△	△	△		■	
RFP Summary	○	●	✖		✖	✖	✖		○	
Stakeholder Management Form	△	●	▲	▲	▲	▲	△		△	
Technical Approach	■	●	△	▲	△	△	△		✖	

Legend
- ■ Must Approve
- ● Specialized Responsibility
- ▲ May Be Consulted
- △ Will Be Consulted
- □ Must Issue
- ✖ Must Review
- ○ Will Receive

Linear responsibility chart

Combined
Worksheet/Bar Chart
Report Produced with the
Aid of Microsoft Project™

ID	WBS	Task Name	Duration	Cost
1	00	Megacorp Project (Best Site)	222 days	$102,675.00
2	M1	Start	0 days	$0.00
3	01	Work Plans	20 days	$9,000.00
4	0101	Investigation	5 days	$3,500.00
5	0102	Health & Safety	5 days	$750.00
6	0103	Quality Assurance	5 days	$3,000.00
7	0104	Internal Editing & Review	10 days	$1,000.00
8	M2	Submit To Customer	0 days	$0.00
9	0105	Customer Review	5 days	$0.00
10	M3	Customer Approval	0 days	$0.00
11	0106	Task Management	20 days	$750.00
12	02	Investigation	56 days	$34,325.00
13	0201	Well Survey & Repair	2 days	$1,125.00
14	0202	Well Gauging & Development	2 days	$1,125.00
15	0203	Slug Tests	1 day	$1,500.00
16	0204	Well Sampling	2 days	$1,125.00
17	0205	Groundwater Analysis	5 days	$5,500.00
18	0206	Geoprobe survey	2 days	$2,000.00
19	0207	Geoprobe Soil Analysis	10 days	$5,500.00
20	0207	New Well Installation	2 days	$3,000.00
21	0208	Soil Analysis (New Wells)	5 days	$1,800.00
22	0209	Gauge & Develop New Wells	1 day	$575.00
23	0210	Sample New Wells	1 day	$575.00
24	0211	Groundwater Analysis	5 days	$600.00
25	0212	Investigation Report	15 days	$6,500.00
26	0213	Report Editing & Review	5 days	$1,000.00

ID	WBS	Task Name	Duration	Cost	3/1	March 3/8	3/15	3/22	3/29	4/5	April 4/12	4/19
27	M4	Submit to Customer	0 days	$0.00								
28	0214	Customer Review	10 days	$0.00								
29	M5	Customer Approval	0 days	$0.00								
30	0215	Investigation Task Management	55 days	$2,400.00								
31	03	**Preliminary Risk Assessment**	**20 days**	**$8,600.00**								
32	0301	Estimate Exposure Point Conc.	2 days	$2,000.00								
33	0302	Risk Characterization	2 days	$1,500.00								
34	0303	Develop Clean-up Goals	2 days	$1,500.00								
35	0304	Risk Assessment Report	3 days	$2,000.00								
36	0305	Internal Review & Edit	1 day	$1,000.00								
37	M6	Submit To Customer	0 days	$0.00								
38	0306	Customer Review	10 days	$0.00								
39	M7	Customer Approval	0 days	$0.00								
40	0307	Task Management	20 days	$600.00								
41	04	**Focused Feasibility Study**	**15 days**	**$5,300.00**								
42	0401	Develop Alternatives	1 day	$600.00								
43	0402	Screen & Analyze Alternative	2 days	$1,200.00								
44	0403	Report Development	5 days	$2,500.00								
45	0404	Feasibility Report Edit & Review	2 days	$600.00								
46	M8	Submit To Customer	0 days	$0.00								
47	0405	Customer Review	5 days	$0.00								
48	M9	Customer Approval	0 days	$0.00								
49	0407	Feasibility Study Task Management	15 days	$400.00								
50	05	**Pilot Test**	**56 days**	**$22,450.00**								
51	0501	**Pilot Test Work Plan**	**18 days**	**$3,100.00**								
52	050101	Draft Plan	10 days	$2,000.00								

ID	WBS	Task Name	Duration	Cost	March 3/1	3/8	3/15	3/22	3/29	April 4/5	4/12	4/19
53	050102	Revise Health & Safety Plan	1 day	$350.00								
54	050103	Work Plan Edit & Review	2 days	$750.00								
55	M10	Submit to Customer	0 days	$0.00								
56	050104	Customer Review	5 days	$0.00								
57	M11	Customer Approval	0 days	$0.00								
58	0502	Pilot Test Execution	21 days	$14,000.00								
59	050201	Order Equipment & Materials	2 days	$5,000.00								
60	050202	Inspect/Stock Eq. & Materials	10 days	$1,000.00								
61	050203	Install Test Wells	2 days	$2,500.00								
62	050204	Shakedown Equipment	5 days	$2,000.00								
63	050205	Conduct Test	2 days	$1,000.00								
64	050206	Pilot Test Report	2 days	$2,500.00								
65	0503	Interpret Data	17 days	$3,750.00								
66	050301	Draft Report	5 days	$1,000.00								
67	050302	Internal Review & Edit	5 days	$2,000.00								
68	050303	Submit To Customer	2 days	$750.00								
69	M12	Customer Review	0 days	$0.00								
70	050304	Customer Approval	5 days	$0.00								
71	M13	Task Management	0 days	$0.00								
72	0504	Design	56 days	$1,600.00								
73	06	Engineering Analysis	55 days	$17,000.00								
74	0601	Mechanical Drawings	5 days	$1,200.00								
75	0602	Site Drawings	5 days	$1,500.00								
76	0603	Instrumentation & Controls	5 days	$1,500.00								
77	0604	Structural Drawings	5 days	$2,000.00								
78	0605		5 days	$1,000.00								

ID	WBS	Task Name	Duration	Cost	March					April		
					3/1	3/8	3/15	3/22	3/29	4/5	4/12	4/19
79	0606	Electrical Drawings	5 days	$1,200.00								
80	0607	Equipment Specifications	5 days	$1,500.00								
81	0608	Construction Specifications	5 days	$1,000.00								
82	0609	Construction Cost Estimates	5 days	$1,200.00								
83	0610	Final Design Report	10 days	$2,500.00								
84	0611	Internal Review & Edit	5 days	$1,200.00								
85	M14	Submit to Customer	0 days	$0.00								
86	0612	Customer Review	15 days	$0.00								
87	M15	Customer Approval	0 days	$0.00								
88	0613	Task Management	55 days	$1,200.00								
89	06	Project Management	221 days	$6,000.00								
90	M16	End	0 days	$0.00								

ID	WBS	Task Name	Duration	Cost	April		May				June	
					4/26	5/3	5/10	5/17	5/24	5/31	6/7	6/14
1	00	Megacorp Project (Best Site)	222 days	$102,675.00								
2	M1	Start	0 days	$0.00								
3	01	Work Plans	20 days	$9,000.00								
4	0101	Investigation	5 days	$3,500.00								
5	0102	Health & Safety	5 days	$750.00								
6	0103	Quality Assurance	5 days	$3,000.00								
7	0104	Internal Editing & Review	10 days	$1,000.00								
8	M2	Submit To Customer	0 days	$0.00								
9	0105	Customer Review	5 days	$0.00								
10	M3	Customer Approval	0 days	$0.00								
11	0106	Task Management	20 days	$750.00								
12	02	Investigation	56 days	$34,325.00								
13	0201	Well Survey & Repair	2 days	$1,125.00								
14	0202	Well Gauging & Development	2 days	$1,125.00								
15	0203	Slug Tests	1 day	$1,500.00								
16	0204	Well Sampling	2 days	$1,125.00								
17	0205	Groundwater Analysis	5 days	$5,500.00								
18	0206	Geoprobe survey	2 days	$2,000.00								
19	0207	Geoprobe Soil Analysis	10 days	$5,500.00								
20	0207	New Well Installation	2 days	$3,000.00								
21	0208	Soil Analysis (New Wells)	5 days	$1,800.00								
22	0209	Gauge & Develop New Wells	1 day	$575.00								
23	0210	Sample New Wells	1 day	$575.00								
24	0211	Groundwater Analysis	5 days	$600.00								
25	0212	Investigation Report	15 days	$6,500.00								
26	0213	Report Editing & Review	5 days	$1,000.00								

ID	WBS	Task Name	Duration	Cost	April 4/26	May 5/3	5/10	5/17	5/24	5/31	June 6/7	6/14
27	M4	Submit to Customer	0 days	$0.00								
28	0214	Customer Review	10 days	$0.00								
29	M5	Customer Approval	0 days	$0.00								
30	0215	Investigation Task Management	55 days	$2,400.00								
31	03	**Preliminary Risk Assessment**	**20 days**	**$8,600.00**								
32	0301	Estimate Exposure Point Conc.	2 days	$2,000.00								
33	0302	Risk Characterization	2 days	$1,500.00								
34	0303	Develop Clean-up Goals	2 days	$1,500.00								
35	0304	Risk Assessment Report	3 days	$2,000.00								
36	0305	Internal Review & Edit	1 day	$1,000.00								
37	M6	Submit To Customer	0 days	$0.00								
38	0306	Customer Review	10 days	$0.00								
39	M7	Customer Approval	0 days	$0.00								
40	0307	Task Management	20 days	$600.00								
41	04	**Focused Feasibility Study**	**15 days**	**$5,300.00**								
42	0401	Develop Alternatives	1 day	$600.00								
43	0402	Screen & Analyze Alternative	2 days	$1,200.00								
44	0403	Report Development	5 days	$2,500.00								
45	0404	Feasibility Report Edit & Review	2 days	$600.00								
46	M8	Submit To Customer	0 days	$0.00								
47	0405	Customer Review	5 days	$0.00								
48	M9	Customer Approval	0 days	$0.00								
49	0407	Feasibility Study Task Management	15 days	$400.00								
50	05	**Pilot Test**	**56 days**	**$22,450.00**								
51	0501	**Pilot Test Work Plan**	**18 days**	**$3,100.00**								
52	050101	Draft Plan	10 days	$2,000.00								

ID	WBS	Task Name	Duration	Cost	April 4/26	5/3	5/10	May 5/17	5/24	5/31	June 6/7	6/14
53	050102	Revise Health & Safety Plan	1 day	$350.00								
54	050103	Work Plan Edit & Review	2 days	$750.00								
55	M10	Submit to Customer	0 days	$0.00								
56	050104	Customer Review	5 days	$0.00								
57	M11	Customer Approval	0 days	$0.00								
58	0502	**Pilot Test Execution**	**21 days**	**$14,000.00**								
59	050201	Order Equipment & Materials	2 days	$5,000.00								
60	050202	Inspect/Stock Eq. & Materials	10 days	$1,000.00								
61	050203	Install Test Wells	2 days	$2,500.00								
62	050204	Shakedown Equipment	5 days	$2,000.00								
63	050205	Conduct Test	2 days	$1,000.00								
64	050206	**Pilot Test Report**	2 days	$2,500.00								
65	0503	**Interpret Data**	**17 days**	**$3,750.00**								
66	050301	Draft Report	5 days	$1,000.00								
67	050302	Internal Review & Edit	5 days	$2,000.00								
68	050303	Submit To Customer	2 days	$750.00								
69	M12	Customer Review	0 days	$0.00								
70	050304	Customer Approval	5 days	$0.00								
71	M13	**Task Management**	0 days	$0.00								
72	0504	**Design**	56 days	$1,600.00								
73	06	**Engineering Analysis**	**55 days**	**$17,000.00**								
74	0601	Mechanical Drawings	5 days	$1,200.00								
75	0602	Site Drawings	5 days	$1,500.00								
76	0603	Instrumentation & Controls	5 days	$1,500.00								
77	0604	Structural Drawings	5 days	$2,000.00								
78	0605	Site Drawings	5 days	$1,000.00								

ID	WBS	Task Name	Duration	Cost	April		May					June	
					4/26	5/3	5/10	5/17	5/24	5/31	6/7	6/14	
79	0606	Electrical Drawings	5 days	$1,200.00								
80	0607	Equipment Specifications	5 days	$1,500.00								
81	0608	Construction Specifications	5 days	$1,000.00								
82	0609	Construction Cost Estimates	5 days	$1,200.00								
83	0610	Final Design Report	10 days	$2,500.00								
84	0611	Internal Review & Edit	5 days	$1,200.00								
85	M14	Submit to Customer	0 days	$0.00								
86	0612	Customer Review	15 days	$0.00								
87	M15	Customer Approval	0 days	$0.00								
88	0613	Task Management	55 days	$1,200.00								
89	06	Project Management	221 days	$6,000.00								
90	M16	End	0 days	$0.00								

ID	WBS	Task Name	Duration	Cost	June 6/21	6/28	7/5	July 7/12	7/19	7/26	August 8/2	8/9
1	00	Megacorp Project (Best Site)	222 days	$102,675.00								
2	M1	Start	0 days	$0.00								
3	01	Work Plans	20 days	$9,000.00								
4	0101	Investigation	5 days	$3,500.00								
5	0102	Health & Safety	5 days	$750.00								
6	0103	Quality Assurance	5 days	$3,000.00								
7	0104	Internal Editing & Review	10 days	$1,000.00								
8	M2	Submit To Customer	0 days	$0.00								
9	0105	Customer Review	5 days	$0.00								
10	M3	Customer Approval	0 days	$0.00								
11	0106	Task Management	20 days	$750.00								
12	02	Investigation	56 days	$34,325.00								
13	0201	Well Survey & Repair	2 days	$1,125.00								
14	0202	Well Gauging & Development	2 days	$1,125.00								
15	0203	Slug Tests	1 day	$1,500.00								
16	0204	Well Sampling	2 days	$1,125.00								
17	0205	Groundwater Analysis	5 days	$5,500.00								
18	0206	Geoprobe survey	2 days	$2,000.00								
19	0207	Geoprobe Soil Analysis	10 days	$5,500.00								
20	0207	New Well Installation	2 days	$3,000.00								
21	0208	Soil Analysis (New Wells)	5 days	$1,800.00								
22	0209	Gauge & Develop New Wells	1 day	$575.00								
23	0210	Sample New Wells	1 day	$575.00								
24	0211	Groundwater Analysis	5 days	$600.00								
25	0212	Investigation Report	15 days	$6,500.00								
26	0213	Report Editing & Review	5 days	$1,000.00								

ID	WBS	Task Name	Duration	Cost
27	M4	Submit to Customer	0 days	$0.00
28	0214	Customer Review	10 days	$0.00
29	M5	Customer Approval	0 days	$0.00
30	0215	Investigation Task Management	55 days	$2,400.00
31	03	**Preliminary Risk Assessment**	**20 days**	**$8,600.00**
32	0301	Estimate Exposure Point Conc.	2 days	$2,000.00
33	0302	Risk Characterization	2 days	$1,500.00
34	0303	Develop Clean-up Goals	2 days	$1,500.00
35	0304	Risk Assessment Report	3 days	$2,000.00
36	0305	Internal Review & Edit	1 day	$1,000.00
37	M6	Submit To Customer	0 days	$0.00
38	0306	Customer Review	10 days	$0.00
39	M7	Customer Approval	0 days	$0.00
40	0307	Task Management	20 days	$600.00
41	04	**Focused Feasibility Study**	**15 days**	**$5,300.00**
42	0401	Develop Alternatives	1 day	$600.00
43	0402	Screen & Analyze Alternative	2 days	$1,200.00
44	0403	Report Development	5 days	$2,500.00
45	0404	Feasibility Report Edit & Review	2 days	$600.00
46	M8	Submit To Customer	0 days	$0.00
47	0405	Customer Review	5 days	$0.00
48	M9	Customer Approval	0 days	$0.00
49	0407	Feasibility Study Task Management	15 days	$400.00
50	05	**Pilot Test**	**56 days**	**$22,450.00**
51	0501	**Pilot Test Work Plan**	**18 days**	**$3,100.00**
52	050101	Draft Plan	10 days	$2,000.00

ID	WBS	Task Name	Duration	Cost	June 6/21	6/28	7/5	July 7/12	7/19	7/26	August 8/2	8/9
53	050102	Revise Health & Safety Plan	1 day	$350.00								
54	050103	Work Plan Edit & Review	2 days	$750.00								
55	M10	Submit to Customer	0 days	$0.00								
56	050104	Customer Review	5 days	$0.00								
57	M11	Customer Approval	0 days	$0.00								
58	0502	Pilot Test Execution	21 days	$14,000.00								
59	050201	Order Equipment & Materials	2 days	$5,000.00								
60	050202	Inspect/Stock Eq. & Materials	10 days	$1,000.00								
61	050203	Install Test Wells	2 days	$2,500.00								
62	050204	Shakedown Equipment	5 days	$2,000.00								
63	050205	Conduct Test	2 days	$1,000.00								
64	050206	Pilot Test Report	2 days	$2,500.00								
65	0503	Interpret Data	17 days	$3,750.00								
66	050301	Draft Report	5 days	$1,000.00								
67	050302	Internal Review & Edit	5 days	$2,000.00								
68	050303	Submit To Customer	2 days	$750.00								
69	M12	Customer Review	0 days	$0.00								
70	050304	Customer Approval	5 days	$0.00								
71	M13	Task Management	0 days	$0.00								
72	0504	Design	56 days	$1,600.00								
73	06	Engineering Analysis	55 days	$17,000.00								
74	0601	Mechanical Drawings	5 days	$1,200.00								
75	0602	Site Drawings	5 days	$1,500.00								
76	0603	Instrumentation & Controls	5 days	$1,500.00								
77	0604	Structural Drawings	5 days	$2,000.00								
78	0605		5 days	$1,000.00								

ID	WBS	Task Name	Duration	Cost	June 6/21	6/28	7/5	July 7/12	7/19	7/26	August 8/2	8/9
79	0606	Electrical Drawings	5 days	$1,200.00								
80	0607	Equipment Specifications	5 days	$1,500.00								
81	0608	Construction Specifications	5 days	$1,000.00								
82	0609	Construction Cost Estimates	5 days	$1,200.00								
83	0610	Final Design Report	10 days	$2,500.00								
84	0611	Internal Review & Edit	5 days	$1,200.00								
85	M14	Submit to Customer	0 days	$0.00								
86	0612	Customer Review	15 days	$0.00								
87	M15	Customer Approval	0 days	$0.00								
88	0613	Task Management	55 days	$1,200.00								
89	06	Project Management	221 days	$6,000.00								
90	M16	End	0 days	$0.00								

ID	WBS	Task Name	Duration	Cost	August 8/16	8/23	8/30	September 9/6	9/13	9/20	9/27	October 10/4
1	00	Megacorp Project (Best Site)	222 days	$102,675.00								
2	M1	Start	0 days	$0.00								
3	01	Work Plans	20 days	$9,000.00								
4	0101	Investigation	5 days	$3,500.00								
5	0102	Health & Safety	5 days	$750.00								
6	0103	Quality Assurance	5 days	$3,000.00								
7	0104	Internal Editing & Review	10 days	$1,000.00								
8	M2	Submit To Customer	0 days	$0.00								
9	0105	Customer Review	5 days	$0.00								
10	M3	Customer Approval	0 days	$0.00								
11	0106	Task Management	20 days	$750.00								
12	02	Investigation	56 days	$34,325.00								
13	0201	Well Survey & Repair	2 days	$1,125.00								
14	0202	Well Gauging & Development	2 days	$1,125.00								
15	0203	Slug Tests	1 day	$1,500.00								
16	0204	Well Sampling	2 days	$1,125.00								
17	0205	Groundwater Analysis	5 days	$5,500.00								
18	0206	Geoprobe survey	2 days	$2,000.00								
19	0207	Geoprobe Soil Analysis	10 days	$5,500.00								
20	0207	New Well Installation	2 days	$3,000.00								
21	0208	Soil Analysis (New Wells)	5 days	$1,800.00								
22	0209	Gauge & Develop New Wells	1 day	$575.00								
23	0210	Sample New Wells	1 day	$575.00								
24	0211	Groundwater Analysis	5 days	$600.00								
25	0212	Investigation Report	15 days	$6,500.00								
26	0213	Report Editing & Review	5 days	$1,000.00								

ID	WBS	Task Name	Duration	Cost
27	M4	Submit to Customer	0 days	$0.00
28	0214	Customer Review	10 days	$0.00
29	M5	Customer Approval	0 days	$0.00
30	0215	Investigation Task Management	55 days	$2,400.00
31	03	**Preliminary Risk Assessment**	**20 days**	**$8,600.00**
32	0301	Estimate Exposure Point Conc.	2 days	$2,000.00
33	0302	Risk Characterization	2 days	$1,500.00
34	0303	Develop Clean-up Goals	2 days	$1,500.00
35	0304	Risk Assessment Report	3 days	$2,000.00
36	0305	Internal Review & Edit	1 day	$1,000.00
37	M6	Submit To Customer	0 days	$0.00
38	0306	Customer Review	10 days	$0.00
39	M7	Customer Approval	0 days	$0.00
40	0307	Task Management	20 days	$600.00
41	04	**Focused Feasibility Study**	**15 days**	**$5,300.00**
42	0401	Develop Alternatives	1 day	$600.00
43	0402	Screen & Analyze Alternative	2 days	$1,200.00
44	0403	Report Development	5 days	$2,500.00
45	0404	Feasibility Report Edit & Review	2 days	$600.00
46	M8	Submit To Customer	0 days	$0.00
47	0405	Customer Review	5 days	$0.00
48	M9	Customer Approval	0 days	$0.00
49	0407	Feasibility Study Task Management	15 days	$400.00
50	05	**Pilot Test**	**56 days**	**$22,450.00**
51	0501	**Pilot Test Work Plan**	**18 days**	**$3,100.00**
52	050101	Draft Plan	10 days	$2,000.00

ID	WBS	Task Name	Duration	Cost
53	050102	Revise Health & Safety Plan	1 day	$350.00
54	050103	Work Plan Edit & Review	2 days	$750.00
55	M10	Submit to Customer	0 days	$0.00
56	050104	Customer Review	5 days	$0.00
57	M11	Customer Approval	0 days	$0.00
58	0502	**Pilot Test Execution**	21 days	$14,000.00
59	050201	Order Equipment & Materials	2 days	$5,000.00
60	050202	Inspect/Stock Eq. & Materials	10 days	$1,000.00
61	050203	Install Test Wells	2 days	$2,500.00
62	050204	Shakedown Equipment	5 days	$2,000.00
63	050205	Conduct Test	2 days	$1,000.00
64	050206	**Pilot Test Report**	2 days	$2,500.00
65	0503	**Interpret Data**	17 days	$3,750.00
66	050301	Draft Report	5 days	$1,000.00
67	050302	Internal Review & Edit	5 days	$2,000.00
68	050303	Submit To Customer	2 days	$750.00
69	M12	Customer Review	0 days	$0.00
70	050304	Customer Approval	5 days	$0.00
71	M13	**Task Management**	0 days	$0.00
72	0504	**Design**	56 days	$1,600.00
73	06	**Engineering Analysis**	55 days	$17,000.00
74	0601	Mechanical Drawings	5 days	$1,200.00
75	0602	Site Drawings	5 days	$1,500.00
76	0603	Instrumentation & Controls	5 days	$1,500.00
77	0604	Structural Drawings	5 days	$2,000.00
78	0605		5 days	$1,000.00

ID	WBS	Task Name	Duration	Cost	August 8/16	8/23	8/30	9/6	September 9/13	9/20	9/27	October 10/4
79	0606	Electrical Drawings	5 days	$1,200.00								
80	0607	Equipment Specifications	5 days	$1,500.00								
81	0608	Construction Specifications	5 days	$1,000.00								
82	0609	Construction Cost Estimates	5 days	$1,200.00								
83	0610	Final Design Report	10 days	$2,500.00								
84	0611	Internal Review & Edit	5 days	$1,200.00								
85	M14	Submit to Customer	0 days	$0.00								
86	0612	Customer Review	15 days	$0.00								
87	M15	Customer Approval	0 days	$0.00								
88	0613	Task Management	55 days	$1,200.00								
89	06	Project Management	221 days	$6,000.00								
90	M16	End	0 days	$0.00								

ID	WBS	Task Name	Duration	Cost	10/11	10/18	10/25	11/1	11/8	11/15	11/22	11/29
1	00	Megacorp Project (Best Site)	222 days	$102,675.00								
2	M1	Start	0 days	$0.00								
3	01	Work Plans	20 days	$9,000.00								
4	0101	Investigation	5 days	$3,500.00								
5	0102	Health & Safety	5 days	$750.00								
6	0103	Quality Assurance	5 days	$3,000.00								
7	0104	Internal Editing & Review	10 days	$1,000.00								
8	M2	Submit To Customer	0 days	$0.00								
9	0105	Customer Review	5 days	$0.00								
10	M3	Customer Approval	0 days	$0.00								
11	0106	Task Management	20 days	$750.00								
12	02	Investigation	56 days	$34,325.00								
13	0201	Well Survey & Repair	2 days	$1,125.00								
14	0202	Well Gauging & Development	2 days	$1,125.00								
15	0203	Slug Tests	1 day	$1,500.00								
16	0204	Well Sampling	2 days	$1,125.00								
17	0205	Groundwater Analysis	5 days	$5,500.00								
18	0206	Geoprobe survey	2 days	$2,000.00								
19	0207	Geoprobe Soil Analysis	10 days	$5,500.00								
20	0207	New Well Installation	2 days	$3,000.00								
21	0208	Soil Analysis (New Wells)	5 days	$1,800.00								
22	0209	Gauge & Develop New Wells	1 day	$575.00								
23	0210	Sample New Wells	1 day	$575.00								
24	0211	Groundwater Analysis	5 days	$600.00								
25	0212	Investigation Report	15 days	$6,500.00								
26	0213	Report Editing & Review	5 days	$1,000.00								

ID	WBS	Task Name	Duration	Cost	October 10/11	10/18	10/25	November 11/1	11/8	11/15	11/22	11/29
27	M4	Submit to Customer	0 days	$0.00								
28	0214	Customer Review	10 days	$0.00								
29	M5	Customer Approval	0 days	$0.00								
30	0215	Investigation Task Management	55 days	$2,400.00								
31	03	**Preliminary Risk Assessment**	**20 days**	**$8,600.00**								
32	0301	Estimate Exposure Point Conc.	2 days	$2,000.00								
33	0302	Risk Characterization	2 days	$1,500.00								
34	0303	Develop Clean-up Goals	2 days	$1,500.00								
35	0304	Risk Assessment Report	3 days	$2,000.00								
36	0305	Internal Review & Edit	1 day	$1,000.00								
37	M6	Submit To Customer	0 days	$0.00								
38	0306	Customer Review	10 days	$0.00								
39	M7	Customer Approval	0 days	$0.00								
40	0307	Task Management	20 days	$600.00								
41	04	**Focused Feasibility Study**	**15 days**	**$5,300.00**								
42	0401	Develop Alternatives	1 day	$600.00								
43	0402	Screen & Analyze Alternative	2 days	$1,200.00								
44	0403	Report Development	5 days	$2,500.00								
45	0404	Feasibility Report Edit & Review	2 days	$600.00								
46	M8	Submit To Customer	0 days	$0.00								
47	0405	Customer Review	5 days	$0.00								
48	M9	Customer Approval	0 days	$0.00								
49	0407	Feasibility Study Task Management	15 days	$400.00								
50	05	**Pilot Test**	**56 days**	**$22,450.00**								
51	0501	**Pilot Test Work Plan**	**18 days**	**$3,100.00**								
52	050101	Draft Plan	10 days	$2,000.00								

ID	WBS	Task Name	Duration	Cost
53	050102	Revise Health & Safety Plan	1 day	$350.00
54	050103	Work Plan Edit & Review	2 days	$750.00
55	M10	Submit to Customer	0 days	$0.00
56	050104	Customer Review	5 days	$0.00
57	M11	Customer Approval	0 days	$0.00
58	0502	Pilot Test Execution	21 days	$14,000.00
59	050201	Order Equipment & Materials	2 days	$5,000.00
60	050202	Inspect/Stock Eq. & Materials	10 days	$1,000.00
61	050203	Install Test Wells	2 days	$2,500.00
62	050204	Shakedown Equipment	5 days	$2,000.00
63	050205	Conduct Test	2 days	$1,000.00
64	050206	Pilot Test Report	2 days	$2,500.00
65	0503	Interpret Data	17 days	$3,750.00
66	050301	Draft Report	5 days	$1,000.00
67	050302	Internal Review & Edit	5 days	$2,000.00
68	050303	Submit To Customer	2 days	$750.00
69	M12	Customer Review	0 days	$0.00
70	050304	Customer Approval	5 days	$0.00
71	M13	Task Management	0 days	$0.00
72	0504	Design	56 days	$1,600.00
73	06	Engineering Analysis	55 days	$17,000.00
74	0601	Mechanical Drawings	5 days	$1,200.00
75	0602	Site Drawings	5 days	$1,500.00
76	0603	Instrumentation & Controls	5 days	$1,500.00
77	0604	Structural Drawings	5 days	$2,000.00
78	0605		5 days	$1,000.00

ID	WBS	Task Name	Duration	Cost	October			November				
					10/11	10/18	10/25	11/1	11/8	11/15	11/22	11/29
79	0606	Electrical Drawings	5 days	$1,200.00								
80	0607	Equipment Specifications	5 days	$1,500.00								
81	0608	Construction Specifications	5 days	$1,000.00								
82	0609	Construction Cost Estimates	5 days	$1,200.00								
83	0610	Final Design Report	10 days	$2,500.00								
84	0611	Internal Review & Edit	5 days	$1,200.00								
85	M14	Submit to Customer	0 days	$0.00								
86	0612	Customer Review	15 days	$0.00								
87	M15	Customer Approval	0 days	$0.00								
88	0613	Task Management	55 days	$1,200.00								
89	06	Project Management	221 days	$6,000.00								
90	M16	End	0 days	$0.00								

ID	WBS	Task Name	Duration	Cost	December 12/6	12/13	12/20	12/27	1/3	January 1/10	1/17	1/24
1	00	**Megacorp Project (Best Site)**	**222 days**	**$102,675.00**								
2	M1	**Start**	0 days	$0.00								
3	01	**Work Plans**	**20 days**	**$9,000.00**								
4	0101	Investigation	5 days	$3,500.00								
5	0102	Health & Safety	5 days	$750.00								
6	0103	Quality Assurance	5 days	$3,000.00								
7	0104	Internal Editing & Review	10 days	$1,000.00								
8	M2	Submit To Customer	0 days	$0.00								
9	0105	Customer Review	5 days	$0.00								
10	M3	Customer Approval	0 days	$0.00								
11	0106	Task Management	20 days	$750.00								
12	02	**Investigation**	**56 days**	**$34,325.00**								
13	0201	Well Survey & Repair	2 days	$1,125.00								
14	0202	Well Gauging & Development	2 days	$1,125.00								
15	0203	Slug Tests	1 day	$1,500.00								
16	0204	Well Sampling	2 days	$1,125.00								
17	0205	Groundwater Analysis	5 days	$5,500.00								
18	0206	Geoprobe survey	2 days	$2,000.00								
19	0207	Geoprobe Soil Analysis	10 days	$5,500.00								
20	0207	New Well Installation	2 days	$3,000.00								
21	0208	Soil Analysis (New Wells)	5 days	$1,800.00								
22	0209	Gauge & Develop New Wells	1 day	$575.00								
23	0210	Sample New Wells	1 day	$575.00								
24	0211	Groundwater Analysis	5 days	$600.00								
25	0212	Investigation Report	15 days	$6,500.00								
26	0213	Report Editing & Review	5 days	$1,000.00								

ID	WBS	Task Name	Duration	Cost	12/6	December 12/13	12/20	12/27	1/3	January 1/10	1/17	1/24
27	M4	Submit to Customer	0 days	$0.00								
28	0214	Customer Review	10 days	$0.00								
29	M5	Customer Approval	0 days	$0.00								
30	0215	Investigation Task Management	55 days	$2,400.00								
31	03	**Preliminary Risk Assessment**	**20 days**	**$8,600.00**								
32	0301	Estimate Exposure Point Conc.	2 days	$2,000.00								
33	0302	Risk Characterization	2 days	$1,500.00								
34	0303	Develop Clean-up Goals	2 days	$1,500.00								
35	0304	Risk Assessment Report	3 days	$2,000.00								
36	0305	Internal Review & Edit	1 day	$1,000.00								
37	M6	Submit To Customer	0 days	$0.00								
38	0306	Customer Review	10 days	$0.00								
39	M7	Customer Approval	0 days	$0.00								
40	0307	Task Management	20 days	$600.00								
41	04	**Focused Feasibility Study**	**15 days**	**$5,300.00**								
42	0401	Develop Alternatives	1 day	$600.00								
43	0402	Screen & Analyze Alternative	2 days	$1,200.00								
44	0403	Report Development	5 days	$2,500.00								
45	0404	Feasibility Report Edit & Review	2 days	$600.00								
46	M8	Submit To Customer	0 days	$0.00								
47	0405	Customer Review	5 days	$0.00								
48	M9	Customer Approval	0 days	$0.00								
49	0407	Feasibility Study Task Management	15 days	$400.00								
50	05	**Pilot Test**	**56 days**	**$22,450.00**								
51	0501	**Pilot Test Work Plan**	**18 days**	**$3,100.00**								
52	050101	Draft Plan	10 days	$2,000.00								

ID	WBS	Task Name	Duration	Cost	December 12/6	12/13	12/20	12/27	1/3	January 1/10	1/17	1/24
53	050102	Revise Health & Safety Plan	1 day	$350.00								
54	050103	Work Plan Edit & Review	2 days	$750.00								
55	M10	Submit to Customer	0 days	$0.00								
56	050104	Customer Review	5 days	$0.00								
57	M11	Customer Approval	0 days	$0.00								
58	0502	**Pilot Test Execution**	21 days	$14,000.00								
59	050201	Order Equipment & Materials	2 days	$5,000.00								
60	050202	Inspect/Stock Eq. & Materials	10 days	$1,000.00								
61	050203	Install Test Wells	2 days	$2,500.00								
62	050204	Shakedown Equipment	5 days	$2,000.00								
63	050205	Conduct Test	2 days	$1,000.00								
64	050206	**Pilot Test Report**	2 days	$2,500.00								
65	0503	**Interpret Data**	17 days	$3,750.00								
66	050301	Draft Report	5 days	$1,000.00								
67	050302	Internal Review & Edit	5 days	$2,000.00								
68	050303	Submit To Customer	2 days	$750.00								
69	M12	Customer Review	0 days	$0.00								
70	050304	Customer Approval	5 days	$0.00								
71	M13	**Task Management**	0 days	$0.00								
72	0504	Design	56 days	$1,600.00								
73	06	**Engineering Analysis**	55 days	$17,000.00								
74	0601	Mechanical Drawings	5 days	$1,200.00								
75	0602	Site Drawings	5 days	$1,500.00								
76	0603	Instrumentation & Controls	5 days	$1,500.00								
77	0604	Structural Drawings	5 days	$2,000.00								
78	0605	Structural Drawings	5 days	$1,000.00								

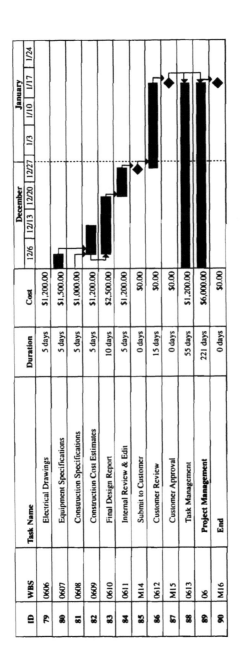

ID	WBS	Task Name	Duration	Cost
79	0606	Electrical Drawings	5 days	$1,200.00
80	0607	Equipment Specifications	5 days	$1,500.00
81	0608	Construction Specifications	5 days	$1,000.00
82	0609	Construction Cost Estimates	5 days	$1,200.00
83	0610	Final Design Report	10 days	$2,500.00
84	0611	Internal Review & Edit	5 days	$1,200.00
85	M14	Submit to Customer	0 days	$0.00
86	0612	Customer Review	15 days	$0.00
87	M15	Customer Approval	0 days	$0.00
88	0613	Task Management	55 days	$1,200.00
89	06	Project Management	221 days	$6,000.00
90	M16	End	0 days	$0.00

Index

9 780367 400217